Praise for *Nature Unbound*

"Read ***Nature Unbound*** and learn the diverse ways in which organized interest groups, and prominent individuals, have sought to impose their idealizations of nature as ecological equilibrium on the rest of us. There is no such thing as nature undisturbed, and bureaucratic bad management is often the unintended consequence of our limited knowledge of ecosystem complexity. Improvement, if attainable, must be more marginal, more decentralized, and focused on learning because no one can know final right answers."

—**Vernon L. Smith**, Nobel Laureate in Economic Sciences; George L. Argyros Endowed Chair in Finance and Economics and Professor of Economics and Law, Chapman University

"In the well-written and highly informative book, ***Nature Unbound***, Simmons, Yonk, and Sim develop the twin concepts of political ecology—the idea that in Washington science is politics—and political entrepreneurship—the notion that every significant political action provides an opportunity for special interest groups to steer the action in their direction. They apply the concepts as they scan the last forty years of the U.S. environmental saga, stopping occasionally to do in-depth analyses of intentions and outcomes. Theirs is not a normative anti-environment cry for deregulation, but rather a carefully reasoned and documented effort to explain how environmental actions based on faulty but popularized notions of science lead inevitably to botched outcomes that fail to redress true environmental concerns. ***Nature Unbound*** should be read, studied and debated by all who take the environment seriously."

—**Bruce Yandle**, Dean Emeritus, College of Business and Behavioral Science, Clemson University; Distinguished Adjunct Professor of Economics, Mercatus Center, George Mason University

"***Nature Unbound*** provides a fascinating look at bureaucracy and environment in the context of a new view of ecology. The new ecology rejects the ideologically based concept of a 'balance of nature' and recognizes variability is fundamental in ecological systems whether or not humans are involved. The book examines the role of politics and entrepreneurship in environmental policy, in the context of the new ecology, and provides an absorbing narration of natural resource legislation, legal activities, and court decisions as well as management policies. The book concludes with five principles for redesigning and incentivizing institutions to be applied to specific individual resource and environmental programs."

—**Roger A. Sedjo**, Senior Fellow, Resources for the Future

"In striving to improve our environmental stewardship, it is important to take off our rose-colored glasses and contemplate the imperfections in our system. In ***Nature Unbound***, Simmons, Yonk, and Sim focus on identifying and explaining deficiencies in long-standing environmental laws. Some readers may find the analysis uncomfortable because it challenges so many deeply ingrained perspectives. Whatever one's view of the authors' criticisms, this book is thought-provoking; it forces us to re-examine the basic incentives and motivations underlying our environmental policies."

—**Gale A. Norton**, former Secretary, U.S. Department of the Interior

"***Nature Unbound*** is a most timely and important book. We are currently in the grip of a revived regulatory utopianism whereby various 'market failures' will be eliminated by diligent, well-informed regulators who have only our best interests at heart. This belief system has recently transformed government policy toward financial services, healthcare—and the environment, which is the subject of this book. In a lively tour through the history of environmental regulation, the authors show how all the elements of the reigning belief are wrong. They tell us how the regulators face information distorted by the interests of the adversaries before them, then how they use the information to pursue personal and bureaucratic interests that ultimately have little to do with improving the environment. I would recommend ***Nature Unbound*** to anyone interested in the environment and especially to those inclined to support further expansion of environmental regulation."

—**Sam Peltzman**, Ralph and Dorothy Keller Distinguished Service Professor Emeritus of Economics, Booth School of Business, University of Chicago; Editor, *Journal of Law and Economics*

"***Nature Unbound*** is a comprehensive discussion of our environmental laws and policies that are conflicted with fallacies and contradictions that cause a reduction in economic output and a damaged natural environment as well. The authors utilize many conceptual tools, but two do much of the intellectual heavy-lifting that drives their analyses and conclusions: 'equilibrium ecology' and 'political entrepreneurship.' This is an important book that hopefully will become influential in producing the legal and legislative improvements needed to increase the environmental quality that will benefit us all."

—**B. Delworth Gardner**, Professor Emeritus of Economics, Brigham Young University

"Environmental policies are ultimately about the human relationship with nature. American environmental thinking about this matter has been characterized by wide intellectual misunderstanding and confusion. Combined with the normal dysfunctions of the American political system, the result has been broad environmental policy failure in the United States. Read the outstanding book *Nature Unbound* to get the details."

  —**Robert H. Nelson**, Professor of Public Policy, University of Maryland; author, *The New Holy Wars: Economic Religion versus Environmental Religion in Contemporary America*

"The U.S. Postal Service is scorned for its inefficiency. Other agencies, such as the EPA, are similarly inept, but it is harder for citizens to observe their poor performance. As the authors explain, we have a 'political ecology.' Politics, not pure environmental science, determines environmental policy. In all areas of the environment—air, water, and land—politicians and bureaucrats play on misguided sympathies to generate support for more of the same. The environment, like society, is dynamic, but government policy is rooted in a mythical environmental notion of static purity. Whether your concern is environmental or economic, *Nature Unbound* is indispensable for anyone to understand the destructive impact of environmental politics."

  —**Roger E. Meiners**, Goolsby Distinguished Professor of Economics and Law, University of Texas, Arlington

"*Nature Unbound* warns that the frequent use of fear-based, emotional science can have the same effect on humans as a deer caught in the headlights. Simmons, Yonk, and Sim offer an escape from this trap and a step forward for Mother Nature. This book is a must read for those seeking an honest and pragmatic path toward improved environmental quality."

  —**Laura E. Huggins**, Chief Executive Officer, Montana Prairie Holdings LLC

"Laws, by definition, tend to lead to litigation. The vaguer the environmental law, the more the various stakeholders are provided with incentives to litigate or procrastinate, rather than even attempt to try to meet the environmental intent of the law. *Nature Unbound* exposes these incentives."

  —**Donald H. Stedman**, John Evans Professor of Chemistry and Biochemistry, University of Denver

"In *Nature Unbound*, Simmons, Yonk, and Sim may have put the final nail in the coffin of the naïve yet abiding notion of the 'balance of nature'—the idea that a place has some unique, ideal natural ecosystem, defined 'out there,' distinct from humans. Commentators since Aldo Leopold have fought this notion, yet it has become more and more entrenched in environmental policy, to the detriment of creation as well as man. So why does a false idea, inevitably leading to bad policy, endure? Using their academic and real-world expertise in politics, Simmons et al. show how groups have perpetuated this notion as a way to promote their own special interests. The book persuasively argues that the 'balance of nature' idea must be replaced by a frank acceptance of the management of nature by humans and—moreover—that it be managed in ways that are robust to special-interest politics. *Nature Unbound* is essential reading in opening our eyes to the reality of present-day environmental failures—and the alternatives."

—**H. Spencer Banzhaf**, Professor of Economics, Andrew Young School of Policy Studies, Georgia State University

"*Nature Unbound* is a revealing and disturbing book. The authors examine the unintended consequences of five major environmental statutes passed by Congress since the 1960s. They outline the ways in which 'political ecology' and 'political entrepreneurship' have channeled the behavior of actors both in and out of government in directions that are opportunistic, costly, and disconnected from legitimate environmental concerns. Simmons, Yonk, and Sim offer principles by which each statute can be amended to more efficiently and effectively address the concerns prompting the initial passage of these laws."

—**John C. Brätland**, Senior Economist, U.S. Department of the Interior

"Environmental policy in the U.S. is based on an incorrect scientific understanding of the 'balance of nature' and the characteristics of a natural equilibrium. Even well-intentioned decision-makers will make incorrect decisions if they start with incorrect premises. Actual decision-makers may be somewhat well-intentioned, but they are also political agents subject to political pressures and special-interest lobbying. In *Nature Unbound*, Simmons, Yonk, and Sim do a very good job of showing how this combination of an incorrect understanding of the world with self-interested decision-makers leads to bad policy, and they have some useful suggestions for improvement. Citizens interested in environmental issues should read this book before advocating further counterproductive policies."

—**Paul H. Rubin**, Samuel Candler Dobbs Professor of Economics, Emory University

"*Nature Unbound* should be required reading for anyone truly interested in policies that will improve environmental quality. It is an insightful and very readable story of why mostly well-intentioned environmental policies have often led to bad outcomes. After carefully documenting the underlying reasons for the lack of environmental success, it offers a set of guiding principles along with constructive and pragmatic suggestions for getting more from key environmental statutes."

>—**Susan E. Dudley**, Director, Regulatory Studies Center, and Distinguished Professor of Practice, Trachtenberg School of Public Policy & Public Administration, George Washington University; former Administrator, Office of Information and Regulatory Affairs, U.S. Office of Management and Budget

"*Nature Unbound* is not only an excellent introduction to the perverse incentives created by the Endangered Species Act and other environmental regulatory schemes, it is a brilliant introduction to the public choice theory of why government so often fails to do what we hope and expect. The book's meticulous scientific evidence, along with its thorough analysis of political failure, make it must-reading for everyone who is concerned about environmental quality as well as for those who want to improve our political system."

>—**Randal O'Toole**, Senior Fellow, Cato Institute

"*Nature Unbound* will be required reading for individuals interested in the protection of endangered species as well as environmental policy more generally. The book is a top priority for scholars, policy-makers, and the general reading public interested in the economic, legal, and political dimensions of environmental policy, drawing from a variety of real world examples to buttress and illustrate the principles of property rights and public choice."

>—**Rodney T. Smith**, Ph.D., President, Stratecon Inc.

"*Nature Unbound* is the best treatment I know of economics and ecology. The application of Austrian and Public Choice insights to environmental policy explains many complex and emotional policies. Environmental sensitivity increases when people become prosperous and well educated. However, environmental issues are scientifically complex and carry heavy emotional baggage. These are ingredients for error and acrimony. Better than anywhere I know, *Nature Unbound* explains the associated rent-seeking, with its lost productive opportunities and environmental damage, and the much needed reforms."

>—**John A. Baden**, Chairman, Foundation for Research on Economics and the Environment

"*Nature Unbound* is a smart, serious book offering crisp thinking about environmental issues that seldom get the critical examination they deserve."

—**Jerry Taylor**, President, Niskanen Center

"*Nature Unbound* will provide the critical, innovative, and careful evaluation so needed to raise issues and encourage debate over environmental issues, which hopefully will lead to more reasoned addressing of environmental concerns."

—**Gary D. Libecap**, Donald Bren Distinguished Professor of Corporate Environmental Management, Bren School of Environmental Science and Management and Department of Economics, University of California, Santa Barbara

"*Nature Unbound* makes a compelling case for abandoning the 'balance of nature' myth and rethinking the environmental laws aimed at maintaining that mythical balance. Simmons, Yonk, and Sim explain why the highly centralized, one-size-fits-all approach to environmental conservation is doomed to failure, and they identify several opportunities for political entrepreneurship that all serious conservationists should consider."

—**Reed Watson**, Executive Director, Property and Environment Research Center

"Nature has often been described as complex and delicate but left to its own devices it will maintain itself in 'balance.' And, when irresponsible people design complex technologies, unexpectedly bad impacts on nature all too often whirl out of control. To address such problems, the responsible people employed by governments have been charged to take care of nature and restore the balances that have been disrupted by the irresponsible rest. Some technologies are banned or strictly regulated, while others are promoted to give 'unbalanced' nature a necessary shove back toward its natural state of 'balance.' Or so much of the green community would have us believe. In *Nature Unbound*, Simmons, Yonk, and Sim challenge this narrative head on and convincingly dismantle it. Time and again, well-intentioned schemes to restore nature to the condition that people have come to believe is natural and pre-ordained by some higher authority just make things worse. This book should be mandatory reading for policymakers, who should have the courage to scrutinize how the environmental laws they promulgate and enforce have worked out and study in depth the many ways in which they have backfired. The rest of us who value the environment as much as they do and marvel at the complex beauty and remarkable resilience of nature should read it too. As *Nature Unbound* convincingly establishes, good intentions do not suffice—to the contrary, in this context all too often they backfire."

—**Peter W. Huber**, Senior Fellow, Manhattan Institute; author, *The Bottomless Well*, *Hard Green*, and *Liability*

Nature Unbound

Other Books by the Authors

Randy T Simmons
Beyond Politics: The Roots of Government Failure
*Critical Thinking About Environmental Issues—
    Endangered Species*

Edited by Shauna Reilly and Ryan M. Yonk
*Direct Democracy in the United States: Petitioners
    as a Reflection of Society*

Randy T Simmons and Ryan M. Yonk and Brian C. Steed
*Green vs. Green: The Political, Legal, and Administrative
    Pitfalls Facing Green Energy Production*

Edited by Randy T Simmons and B. Delworth Gardner
Aquanomics: Water Markets and the Environment

Edited by Terry L. Anderson and Randy T Simmons
*The Political Economy of Customs and Culture:
    Informal Solutions to the Commons Problem*

Edited by Peter F. Galderisi, Michael S. Lyons,
    Randy T Simmons and John Francis
The Politics of Realignment: Party Change in the Mountain West

INDEPENDENT INSTITUTE is a non-profit, non-partisan, public-policy research and educational organization that shapes ideas into profound and lasting impact. The mission of Independent is to boldly advance peaceful, prosperous, and free societies grounded in a commitment to human worth and dignity. Applying independent thinking to issues that matter, we create transformational ideas for today's most pressing social and economic challenges. The results of this work are published as books, our quarterly journal, *The Independent Review*, and other publications and form the basis for numerous conference and media programs. By connecting these ideas with organizations and networks, we seek to inspire action that can unleash an era of unparalleled human flourishing at home and around the globe.

FOUNDER & PRESIDENT
David J. Theroux

RESEARCH DIRECTOR
William F. Shughart II

SENIOR FELLOWS
Bruce L. Benson
Ivan Eland
John C. Goodman
John R. Graham
Robert Higgs
Lawrence J. McQuillan
Robert H. Nelson
Charles V. Peña
Benjamin Powell
William F. Shughart II
Randy T Simmons
Alexander Tabarrok
Alvaro Vargas Llosa
Richard K. Vedder

ACADEMIC ADVISORS
Leszek Balcerowicz
WARSAW SCHOOL OF ECONOMICS

Herman Belz
UNIVERSITY OF MARYLAND

Thomas E. Borcherding
CLAREMONT GRADUATE SCHOOL

Boudewijn Bouckaert
UNIVERSITY OF GHENT, BELGIUM

Allan C. Carlson
HOWARD CENTER

Robert D. Cooter
UNIVERSITY OF CALIFORNIA, BERKELEY

Robert W. Crandall
BROOKINGS INSTITUTION

Richard A. Epstein
NEW YORK UNIVERSITY

B. Delworth Gardner
BRIGHAM YOUNG UNIVERSITY

George Gilder
DISCOVERY INSTITUTE

Nathan Glazer
HARVARD UNIVERSITY

Steve H. Hanke
JOHNS HOPKINS UNIVERSITY

James J. Heckman
UNIVERSITY OF CHICAGO

H. Robert Heller
SONIC AUTOMOTIVE

Deirdre N. McCloskey
UNIVERSITY OF ILLINOIS, CHICAGO

J. Huston McCulloch
OHIO STATE UNIVERSITY

Forrest McDonald
UNIVERSITY OF ALABAMA

Thomas Gale Moore
HOOVER INSTITUTION

Charles Murray
AMERICAN ENTERPRISE INSTITUTE

Michael J. Novak, Jr.
AMERICAN ENTERPRISE INSTITUTE

June E. O'Neill
BARUCH COLLEGE

Charles E. Phelps
UNIVERSITY OF ROCHESTER

Paul H. Rubin
EMORY UNIVERSITY

Bruce M. Russett
YALE UNIVERSITY

Pascal Salin
UNIVERSITY OF PARIS, FRANCE

Vernon L. Smith
CHAPMAN UNIVERSITY

Pablo T. Spiller
UNIVERSITY OF CALIFORNIA, BERKELEY

Joel H. Spring
STATE UNIVERSITY OF NEW YORK, OLD WESTBURY

Richard L. Stroup
NORTH CAROLINA STATE UNIVERSITY

Robert D. Tollison
CLEMSON UNIVERSITY

Arnold S. Trebach
AMERICAN UNIVERSITY

Richard E. Wagner
GEORGE MASON UNIVERSITY

Walter E. Williams
GEORGE MASON UNIVERSITY

Charles Wolf, Jr.
RAND CORPORATION

100 Swan Way, Oakland, California 94621-1428, U.S.A.
Telephone: 510-632-1366 • Facsimile: 510-568-6040 • Email: info@independent.org • www.independent.org

NATURE UNBOUND
Bureaucracy vs. the Environment

Randy T Simmons, Ryan M. Yonk, and Kenneth J. Sim

INDEPENDENT INSTITUTE

OAKLAND, CALIFORNIA

Copyright © 2016 by the Independent Institute

All Rights Reserved. No part of this book may be reproduced or transmitted in any form by electronic or mechanical means now known or to be invented, including photocopying, recording, or information storage and retrieval systems, without permission in writing from the publisher, except by a reviewer who may quote brief passages in a review. Nothing herein should be construed as necessarily reflecting the views of the Institute or as an attempt to aid or hinder the passage of any bill before Congress.

Independent Institute
100 Swan Way, Oakland, CA 94621-1428
Telephone: 510-632-1366
Fax: 510-568-6040
Email: info@independent.org
Website: www.independent.org

Cover Design: Denise Tsui
Cover Image: © Mny-Jhee / Shutterstock

LC CIP Data Available

Contents

1	Politics, Ecology, and Entrepreneurship	1
2	Political Ecology	7
3	Environmental Political Entrepreneurship	21
4	The Politics of Nature	37
5	The Clean Air Act	51
6	The National Environmental Policy Act	73
7	The Clean Water Act	103
8	The Endangered Species Act	127
9	The Wilderness Act	155
10	Renewable Energy Legislation	175
11	Conclusions	187
	Appendix: Federal Land Policy	199
	Notes	215
	References	240
	Index	271
	Acknowledgments	284
	About the Authors	286

1

Politics, Ecology, and Entrepreneurship

WHAT IF WHAT we think we know about ecology and environmental policy is just wrong? What if environmental laws often make things worse? What if the very idea of nature has been hijacked by politics? What if wilderness is something we create in our minds, as opposed to being an actual description of nature? Developing answers to these questions and developing implications of those answers are our purposes in this book. Two themes guide us—political ecology and political entrepreneurship. Combining these two concepts, which we develop in some detail, leads us to recognize that sometimes in their original design and certainly in their implementation, major U.S. environmental laws are more about opportunism and ideology than good management and environmental improvement.

Political Ecology

A well-respected ecologist at Utah State University spent much of his academic life studying landscape changes in and around Yellowstone National Park. When we asked him what ought to be the correct policy for managing the park, anything ranging from actively managing nature to letting nature just happen, a policy known as *natural regulation,* he responded that he does not make policy recommendations; that is for the politicians and agency personnel.

We believe his answer is consistent with what many ecologists would say—their job is to study ecological processes, not to get involved in the politics of management. There is, they believe, somehow a separation between science and politics. Science may inform politics, but they are separate endeavors.

A major theme of this book is that a separation between science and politics is nearly as rare as unicorns. There is a politics of ecology, even though many ecologists refuse to participate in it, at least overtly. This political ecology is the politicization of ecology—not necessarily the *science* of ecology, although it often is—but beliefs about ecology held by the public, press, and policymakers. Those beliefs have become increasingly political. Political ecology underlies the foundational laws of U.S. environmental policy. We will demonstrate that these laws are based on mythology intertwined with naiveté about political processes and false assumptions about environmental processes.

Popular political ecology is based on a flawed understanding of ecology. This misunderstanding is best described as the "balance of nature," which holds that when nature is left undisturbed by humans, it remains in or returns to a perpetual state of balance and harmony. The balance of nature belief asserts that if we "let nature take her course," the environment will take care of itself.

If the premises of the balance of nature belief were correct, then one would expect that allowing nature to take her course in environmental policy-making would lead to the protection and enhancement of biological diversity and ecological integrity for natural resources of the United States. If, however, the balance of nature belief is a political and mythological construct, policies and management based on it will fail to produce the desired results.

The balance of nature belief is romantic and emotionally appealing. It is comforting to believe that nature could maintain and repair herself if only humans would stop meddling in her affairs. Unfortunately, this belief is outdated, has been rejected by modern ecologists, and is simply wrong. Policies based on the balance of nature ideology are ineffective and counterproductive; western environmental philosophies based on these outdated, romantic, even religious assumptions are misleading.[1] That does not change the fact, however, that the balance of nature belief is alive and thriving in the United States among university students, the general public, political activists, and politicians.

Furthermore, balance of nature solutions (also called "steady-state" solutions) to environmental challenges are presently at work throughout all of modern U.S. environmental law and policy. Reading the Endangered Species Act, Clean Water Act, National Environmental Policy Act, or any other major environmental law quickly shows that it is based on equilibrium ecology, which is another term for the balance of nature. Ecologist Norman Christensen, Dean of Duke University's Nicholas School of the Environment, wrote that such laws are based on "the idea that systems tend toward these stable end points and that they are regulated by complex feedbacks—a sort of balance of nature that is almost Aristotelian."[2]

Environmental Political Entrepreneurship

A second major theme that drives our analysis is political entrepreneurship. Environmental preservation is a fundamentally political enterprise. Legislation, regulation, policies, and evaluations are political actions. We suggest that their outcomes might be better understood if we use the lens of political entrepreneurship.

In 2012 we attended public meetings for President Obama's proposed solar energy zones on public lands managed by the U.S. Bureau of Land Management (BLM). The zones would supposedly allow for expedited review and approval of solar energy farms. Opponents of the solar energy zones outnumbered proponents by a wide margin, and some members of the public who were attending the meeting were highly emotional. Most of those who spoke were in favor of solar energy in principle, but not in the particular areas proposed by the Obama administration. Speakers opposed a solar energy zone in the Wah Wah Valley in Utah, for example, because it would destroy pristine views. Dark skies proponents opposed sites in central Arizona because of their possible effects on stargazing. Impacts to desert tortoise habitat and fragile ecosystems were stated as reasons for opposing proposed sites in California. Almost everyone at these meetings claimed to be a stakeholder representing some portion of the public interest. A few people spoke out against the proposed solar sites as representing a silly infatuation with an inefficient and therefore wasteful technology. The BLM employees listened politely.

The conversation at the solar energy zone hearing included references to unspoiled ecosystems, pristine landscapes, the balance of nature, and endangered and threatened species such as the Mojave fringe-toed lizard, flat-tailed horned lizard, golden eagle, desert bighorn, and desert tortoise. Later, the Western Lands Project filed a protest with the U.S. Department of the Interior claiming the solar zones would destroy "one of the last remaining floristic frontiers in the United States" and called the proposal to establish solar energy zones as "taking the least enlightened path possible, while attempting to create the illusion of innovation and progress."[3]

The solar energy zone conversation illustrates how political activists attempt to capture policy processes to achieve their goals. Politicians, political appointees, bureaucrats, and members of organized interest groups evoke emotions and make claims about the public interest to promote their agendas. Paradoxically, their efforts result in legislation and regulation that are often designed (unintentionally or even intentionally) to fail. Sometimes the failure results from legislation and ensuing regulation based on ecologically incorrect assumptions and insufficient knowledge. Alternatively or even concurrently, the legislation is the result of a process that allows narrow interests to trump more general interests. This process is known as political entrepreneurship.

We use the term *entrepreneur* broadly to mean *homo agens* (the human actor) who "possesses the propensity to pursue goals effectively, once ends and means are clearly identified, but also possesses the alertness to identify which ends are to be sought and what means are available."[4] Political entrepreneurial behavior is "alertness to unnoticed opportunities to achieve policy outcomes."[5] For example, an astute political entrepreneur from an environmental advocacy group may claim in a comment on a National Environmental Policy Act document that impacts to a species have been inadequately analyzed. The resulting reanalysis will cost the project money and may delay its implementation until the analysis is deemed acceptable. Thus, the entrepreneur has achieved an outcome (time delay and cost increases) that may please his or her constituency.

Defining political entrepreneurship as alertness to unnoticed opportunities spreads the area of focus from just the study of heroic figures such as John Muir or Theodore Roosevelt to include bureaucrats, political appointees, members of interest groups or think tanks, and interested individuals. Political entrepreneurs can be those who advocate for proposals as well as those who attempt to block them. Political scientist John Kingdon sums up our notion of political entrepreneurship nicely. He said political entrepreneurs "invest their resources — time, energy, reputation, and sometimes money—in the hope of a future return."[6] We would add that the "future return" might be to preserve the status quo.

Just as ecosystems are characterized by competition, disturbance, and succession as organisms attempt to succeed and even thrive, politics is a struggle by competing interest groups and individuals such as Muir, Roosevelt, Gifford Pinchot, and Aldo Leopold to impose their ideals on the rest of society. Political scientist Robert A. Dahl summarized the process years ago: "The making of government policy is not a majestic march of great majorities united upon certain matters of policy. It is the steady appeasement of relatively small groups."[7] "The people" do not create environmental legislation. Organized groups do, sometimes achieving private gains at the public expense. Peter Kareiva, chief scientist for The Nature Conservancy, explained part of what is going on:

> Beneath the invocations of the spiritual and transcendental value of untrammeled nature is an argument for using landscapes for some things and not others: hiking trails rather than roads, science stations rather than logging operations, and hotels for eco-tourists instead of homes. By removing long-established human communities, erecting hotels in their stead, removing unwanted species while supporting more desirable species, drilling wells to water wildlife, and imposing fire management that mixes control with prescribed burns, we create parks that are no less human constructions than Disneyland.[8]

The "Disneylands" created by these political processes come at someone's expense, but seldom at the expense of the political entrepreneurs and their followers.

Implications

Our claims about political ecology are not new. Everyone knows that acts of Congress are political. That is why there are so many lobbyists in Washington, D.C., and state capitals. What is unique in our analysis is the emphasis on politicized or political ecology and entrepreneurship.

Chapters 2 and 3 lay out in some detail the science and politics of political ecology. Some conclusions are startling, disconcerting, and unpopular: stopping logging in the Pacific Northwest will not restore the forests to their pre-Columbian state; saving an endemic species makes little biological sense; nature must be managed; ecosystems (however defined) are not delicately balanced; today's ecosystems did not result from nature "taking its course"; American ecosystems were not in a state of equilibrium when European settlers arrived; and the ecology of our landscapes is more complex and reliant on human intervention than originally thought. The idea that nature has an idealized and balanced state is mythology. Although many scientists no longer promote the balance of nature, environmental organizations still find it useful, and our environmental laws are still predicated on it.

In Chapter 4 we reflect on the politics of defining terms like *nature, wilderness,* and *natural.* "Nature undisturbed" has been a useful myth for political entrepreneurs for many decades. As we will see in more detail throughout other chapters, it justified and animated a long-term political movement to preserve nature, to return much of the United States to an invented state of natural balance, and to create a host of federal laws and subsequent regulations.

Chapters 5 through 9 apply our political ecology analysis to major U.S. environmental legislation: the Clean Air Act, the National Environmental Policy Act, the Clean Water Act, the Endangered Species Act, and the Wilderness Act. Chapter 10 is a case study in the political uses of renewable energy legislation. These chapters review history, assumptions, and especially the opportunities for political entrepreneurship in how legislation was created and then applied.

Chapter 11 lays out some basic ideas for redesigning environmental policies. We recognize that political pressures work against such redesign, but we remain pragmatically hopeful.

Finally, we have included an appendix that reviews the history of public land management. It provides background for Chapters 5 to 9 but, unlike those chapters, does not concentrate on a single legislative act.

2

Political Ecology

IN HIS BOOK about the relationships between humans and nature, ecologist Daniel Botkin explains that ecologists and other environmental scientists reject the balance of nature or steady-state assumption. He claims, however, that "models, theory, management policies, recommendations, etc., in ecology assume that ecosystems and species are in a steady state and will never change, period." He continues,

> If you ask ecologists whether nature is always constant, they will always say, 'No, of course not.' But if you ask them to write down a policy for biological conservation or any kind of environmental management, they will almost always write down a steady-state solution.[1]

Although environmental science may reject the balance of nature ideology[2] (while supporting laws based in it), the general public tends to accept it wholeheartedly. A 2007 study of undergraduate students in the United States showed they "believe this term is descriptive of real ecological systems." In addition, the students continue to believe in the balance of nature even after taking courses in environmental science.[3]

In his book about "reinventing nature," William Cronon claims, "Many popular ideas about the environment are premised on the conviction that nature is a stable, holistic, homeostatic community capable of preserving its natural balance more or less indefinitely if only humans can avoid 'disturbing it.'"[4] This assumption, which he calls "problematic," descends from the work of botanist Frederic Edward Clements, for whom the "landscape is a balance of nature, a steady-state condition maintained so long as every species remains in place."[5]

Central to this belief is the presumption that nature is highly structured, ordered, and regulated and that disturbed ecosystems will return to their original state once the disturbance ceases. This view of nature is an integral part of successional theory, in which species are thought to replace one another in ordered procession culminating in climax communities. It continues to animate many modern activists. On the World Wildlife Fund's website, for example, there is a section titled "Ecological Balance" that contains the following text:

> Ecological balance has been defined by various online dictionaries as "a state of dynamic equilibrium within a community of organisms in which genetic, species and ecosystem diversity remain relatively stable, subject to gradual changes through natural succession" and "A stable balance in the numbers of each species in an ecosystem."

The next paragraph begins, "The most important point being that the natural balance in an ecosystem is maintained."[6]

Rachel Carson, author of *Silent Spring*, is perhaps most responsible for popularizing the idea of a balance of nature. Although she noted that "the balance of nature is not a *status quo*; it is fluid, ever shifting, in a constant state of adjustment," she also claimed that it is no more possible to ignore the balance of nature than a "man perched on the edge of a cliff" can defy the "law of gravity."[7] *Silent Spring* promoted the notion that a delicate and static balance of nature stands in danger of being upset by humans. Carson claimed that it took "eons of time" for life to reach "a state of adjustment and balance with its surroundings."[8]

A belief in a balance of nature gives a strong moral context to environmental protection. From Rachel Carson to Barry Commoner (who famously said, "everything is connected to everything else"), the Club of Rome, Al Gore, and the deep ecologists,[9] there is not only an acceptance that a balance of nature exists but also a belief that upsetting the balance of nature is morally wrong.

This view is not modern, progressive, or scientific. As John Kricher explains in his book, *The Balance of Nature: Ecology's Enduring Myth*, the balance of nature ideology is in reality an old conservative religious view of the natural world that dates to the dawn of written history.[10] Contrary to the constancy and stability of the balance of nature ideology, however, disturbance and change are the normal order in the evolutionary history of the Earth. For example, glaciers have advanced and retreated over the surface of North America repeatedly in the last 3 million years.

Not only has the climate fluctuated widely, but what we in the United States view as normal (that which we have experienced during our lives or since the birth of our nation) is, when viewed from a perspective of the last several hundred thousand years, an abnormally warm, dry period. The normal climate for most of Canada, for instance, is several thousand feet of ice, not what we see today.[11] As one might expect, the distributions of plants and animals have also contracted and expanded over time. Local extinctions are a fact of life, as is the extinction of entire species. Disturbance and change are the only ecosystem constants.

Previously, we noted Daniel Botkin's assertion of a disconnect between the scientific knowledge of ecologists and their policy prescriptions. Such dissonance is not restricted to ecologists but is, we believe, widespread among natural scientists in general, regardless of their disciplines. It is stated clearly by Yellowstone National Park historian Paul Schullery. Although he was writing specifically about aboriginal influences on North American ecosystems, the following statement applies broadly:

> Even among those of us who know and accept this evidence there is a lingering feeling that things were somehow right back then, that some fundamental state of harmony existed between humans and the rest of nature and that North America was a kind of environmental Eden until Europeans arrived.[12]

Lack of Wildlife

One reason for the prevailing popular view of North America as Eden is early accounts that describe uncountable numbers of wildlife prior to widespread European exploration and settlement. Meriwether Lewis and William Clark, for example, recorded herds that were a "moving multitude that darkened the plains from one horizon to the other."[13] Flocks of passenger pigeons were reported to have blocked out the sun for consecutive days.

Historical records, however, do not support the notion that North America once teemed with wildlife. Although some have claimed, for example, that moose in the Rocky Mountains numbered in the tens of thousands before they were slaughtered by unregulated hunting, early fur trappers seldom reported seeing or killing even a single moose.[14] When Peter Skene Ogden's fur brigade killed three moose near present-day Phillipsburg, Montana, in 1825, he noted that it was the first time any of his men had seen a moose in their total of nearly 300 man-years in what is now the western United States during the early 1800s.[15]

Although not as rare as moose, elk were also historically uncommon in the Rocky Mountains. Between 1835 and 1872, for example, twenty different parties spent a total of 765 days traveling through Yellowstone on foot or horseback, yet reported seeing elk only once every eighteen days—in 1995, prior to gray wolf reintroduction, there were nearly 100,000 elk in that ecosystem.[16] The same was true in the Canadian Rockies, where early explorers reported seeing elk only once every thirty-one days.[17]

Moreover, mule and white-tailed deer, antelope, and bighorn sheep were also generally rare or absent when the Rocky Mountains were first visited by Europeans. Accounts of starvation and killing horses for food are common in early journals.[18] The Lewis and Clark expedition ate 190 dogs and 12 horses in areas where wild game was not available.

Except for Idaho's Snake River plains and adjacent areas, few bison were ever seen west of the mountains. The 2013 bison count in Yellowstone National Park estimated there are 4,600 bison, but between 1835 and 1872, early explorers encountered bison only three times during 765 days in the ecosystem.[19] The Columbia Plateau and the Great Basin were practically devoid of game at historical contact.[20] How can we square that historical record with Lewis and Clark's accounts of massive herds of bison and flocks of countless passenger pigeons?

There are at least two possible explanations. First, the bison herds and pigeon flocks were freak population events caused by a sudden disappearance of millions of Native Americans. Second, the bison herds in particular were primarily found in buffer zones between competing tribes, where hunting pressure was low.

Writers have long recognized that Native Americans lacked immunological resistance to epidemic and endemic European diseases and that many epidemics reduced aboriginal numbers by 50 to 90 percent at each passing.[21] Anthropologist Henry Dobyns postulated that Native American populations were severely reduced a hundred years before the Pilgrims set foot at Plymouth Rock.[22] Professor Ann Ramenofsky, who tested Dobyn's hypothesis against the archaeological record, found that the tribes along the middle Missouri River were decimated by European disease ca. 1600 AD, 200 years before the arrival of Lewis and Clark, who were the first Europeans to leave a detailed written record of exploration in the western United States.[23] Taking this factor into consideration, several authors have revised aboriginal population estimates for North America upward to 100 million or more. In 1492 more than 2 million native people may have lived in California alone, and at least 1 million natives occupied the Hawaiian Islands.[24] Native populations that depended on agriculture in the eastern United States and in the Southwest were particularly dense.[25] Moreover, aboriginal people had significant impacts on the environment, often stripping forests for fuel or clearing land for crops that, in turn, increased soil erosion (Diamond, 1988).[26]

These Native American populations were decimated by early contact with Eurasian diseases including tuberculosis, typhoid, diphtheria, smallpox, whooping cough, influenza, yellow fever, scarlet fever, and measles. This disease holocaust swept through native peoples and, according to some estimates, killed 95 percent of them. Two results of this depopulation were (1) wildlife no longer competed with Native Americans for food, and (2) most hunting ceased.

Lack of Competition

In his book *1491: New Revelations of the Americas Before Columbus*, Charles C. Mann cites work by archaeologist Thomas W. Neumann showing almost no pigeon bones were found in archaeological sites.[27] There are bones from many other birds and even fish, but few from passenger pigeons. That is odd if there were in fact huge flocks of passenger pigeons historically. The pigeons roosted in large flocks and were famously dumb—they could be knocked from trees with poles, for example. Mann cites Neumann as saying, "If they are so easy to hunt and you expect people to minimize labor and maximize return, you should have archeological sites just filled with these things. Well you don't."[28] Neumann concludes the pigeons were simply not numerous before Europeans arrived. After disease killed off many if not most of the humans who had been competing with the pigeons for mast (nuts of all kinds), the pigeon populations exploded. He calls the huge flocks observed by Europeans, "outbreak populations—always a symptom of an extraordinarily disrupted ecological system."[29]

The bison story is similar to the pigeon story. Hernando de Soto's expedition spent four years in the Southwest and did not record seeing even one bison, but they recorded seeing a land heavily populated with people. One hundred years later, the French explorer LaSalle saw almost no people but huge bison herds. In his classic book on the history of bison, biologist Valerius Geist wrote, "The post-Columbian abundance of bison in North America, however, was almost certainly due to the decimation of a large portion of the Native American's population by Eurasian diseases that decreased human hunting of bison."[30]

Buffer Zones

Lewis and Clark and others did see large herds of animals sometimes and nearly starved at other times. One reason is likely the existence of aboriginal buffer zones. In a paper, "The Virginia Deer and Intertribal Buffer Zones in the Upper Mississippi Valley," Hickerson explains,

> Warfare between members of the two tribes had the effect of preventing hunters from occupying the best game region intensively enough to deplete the (deer) supply . . . In the one instance in which a lengthy truce was maintained between certain Chippewas and Sioux, the buffer, in effect a protective zone for the deer, was destroyed and famine ensued.[31]

Our colleague Charles Kay uncovered frequent references to buffer zones created by Native American hunting.[32] He found the following in Lewis and Clark's journals: "With regard to game in general, we observe that the greatest quantities of wild animals are usually found in the country lying between two nations at war."[33] In 1859 General Raynolds, who led an expedition across the

Dakotas and Montana, found an abundance of grass but no game east of the Powder River. Along the Powder River, however, he reported an abundance of game and little grass, whereas to the west he again encountered an abundance of grass and no game. Raynolds noted that,

> The presence of these animals [bison] in such large numbers in this barren region [Powder River] is explained by the fact that this valley is a species of neutral ground between the Sioux and the Crows and other bands nearer the mountains, or, more correctly speaking, the common war ground visited only by war parties, who never disturb the game, as they would thereby give notice to their enemies of their presence. For this reason, the buffalo remain here undisturbed and indeed would seem to make the valley a place of refuge.[34]

Tribal territory boundary zones explain how early explorers could encounter an abundance of game in a few locations and a lack of game everywhere else. Many aboriginal buffer zones were up to a hundred miles wide or more.

The Myth of Untouched Wilderness

Clearly, North America was not a wilderness waiting to be discovered. Rather, it was home to tens of millions of aboriginal peoples before the arrival of Europeans and their diseases. Wilderness was not a concept understood or used by native peoples; no native language even contains a word for *wilderness*.[35]

Africa's Serengeti is often used as an example of how North America's prairies must have looked before Europeans supposedly despoiled them. Today's Serengeti, however, is not a true representation of wilderness in Africa. It is a romantic, European view of how "primitive" Africa should have looked.[36] One of the first things Europeans did was remove all of the Serengeti's indigenous peoples, even though there had been hominoid predators in Africa for at least 3.8 million years and despite the fact that our species, *Homo sapiens*, evolved in Africa approximately 100,000 years ago. There is nothing more unnatural than an African ecological system without hominoid predators. The Serengeti, therefore, is not a natural ecosystem nor is it an example of how things were in North America before the arrival of Columbus.

Most national parks, wilderness areas, and nature reserves, are supposedly managed to represent the conditions that existed in pre-Columbian times (i.e., so-called natural or pristine conditions). The National Park Service has often called the northern portion of Yellowstone National Park "America's Serengeti" and in official publications claimed that the current high populations of elk prior to wolf reintroduction approximated the numbers there "before significant human intervention began in the 19th century."[37] Apparently, native people's interventions were not significant, according to the Park Service.

But what is natural? If Native Americans determined the structure of entire plant and animal communities by burning vegetation and limiting wildlife numbers, then they created completely different situations than what we have today. A let-nature-take-its-course approach by modern land managers will not duplicate the ecological conditions under which those communities developed. Because aboriginal predation and burning created those communities, we believe they will be best maintained by duplicating aboriginal influences and processes. Furthermore, unless the importance of aboriginal land management is recognized and modern management practices changed accordingly, we fear our ecosystems will continue to lose the biological diversity and ecological integrity they once had.

As paradoxical as it may sound, nature has to be managed. Setting aside areas as wilderness has been suggested as one way to protect various endangered species. However, this will not preserve a remnant of the past. It will instead create conditions that have not existed for the last 10,000 years. North Americans, for instance, view the Amazon River Basin as a wilderness to be saved and protected. To indigenous peoples, however, the Amazon is a home—a home they have modified to suit human needs by burning vegetation, planting crops, and harvesting game.[38]

Misleading Environmental Myths

In addition to the outdated assumptions described above, additional environmental myths are often invoked. These include the belief that ecosystems are highly evolved and delicately balanced and that old-growth forests were the norm in pre-European times.

Ecosystem Myths

Ecosystems exist only in the mind's eye; they can be defined as anything from a drop of pond water to the entire planet; thus, an ecosystem is whatever you want it to be.[39] Ecosystems are not natural biological units like species; they are artificial constructs of the human imagination. What we commonly call an ecosystem is almost invariably some geographic unit, like the Greater Yellowstone Ecosystem, the Northern Rockies Ecosystem, or some other area that a group or groups are interested in saving, conserving, protecting, or preserving. And in many cases, an endangered species, like the northern spotted owl, the grizzly bear, or the wolf, is at the center of the debate to preserve one geographic area or another. Defining these areas as ecosystems provides emotional content to the debate.

In their campaigns to define and defend ecosystems, environmental activists often claim ecosystems are highly evolved and delicately balanced. But the most elementary classes in evolutionary ecology teach that ecosystems do *not* evolve. Individuals live and die, but only species evolve. Ecosystems generally become more complex over time, as new species arrive or as new species evolve, but by no stretch of logic can it be said that ecosystems evolve. Instead of metaphysical units, ecosystems are little more than semi-random collections of individual species.[40]

In addition, most ecosystems are not delicately balanced, especially systems with large numbers of species in all trophic levels (i.e., primary producers or plants, herbivores, and predators). Most ecosystems are highly redundant. When one species is lost, it is replaced by increases in other similar species.[41] Contrary to what is often claimed, even species that by their numbers are major components of an ecosystem can be lost without a total collapse of the system. This is termed *resiliency*, and most ecosystems are highly resilient.

In pre-Columbian times and early historical times as well, the American chestnut was a major component of eastern deciduous forests. There are closely related species of chestnuts in Asia and Europe. In addition to transporting human diseases across the Atlantic, Europeans brought various plant diseases, too—one of which attacks chestnut trees, the Asian chestnut blight. Although Asian chestnuts had evolved some resistance to this pathogen, the American species had not. So, when chestnut blight accidentally crossed the ocean in 1904, within fifty years it destroyed North America's chestnut trees in a fashion similar to what European diseases did to Native Americans. Today, there are no American chestnut trees. Although the species survives as root sprouts, the disease repeatedly kills the stems before they can grow into trees. Even without chestnuts, however, there still is an eastern deciduous forest. Other species have simply grown up to replace the chestnuts.[42] Similarly, the grizzly bear and gray wolf are gone from most of the western United States, but we still have ecosystems in California, Arizona, Utah, and other places where those species were once found. As Canadian ecologist E.C. Pielou noted,

> The notion espoused by so many nonprofessional ecologists—that the living world is "marvelously" and "delicately" attuned to its environment [balanced]—is not so much a scientifically reasonable theory as a mystically satisfying dogma. Its abandonment might lead to a useful fresh start in environmental politics.[43]

Old-Growth Myths

A popular and enduring vision of America at the time of European contact is that the forests were climax, old-growth, steady-state forests. Not only is this a myth, but it ignores the actions of indigenous peoples. It is often claimed, for instance, that the area that is now the eastern United States was blanketed in climax deciduous forests before Europeans landed. Early accounts depict a forest of widely spaced trees with little understory, a park-like forest through which one could easily ride a horse or drive a wagon. That is because Native Americans had actually changed the forests to grow that way. What European settlers saw as "natural"[44] was in fact the result of significant human influence. The tangled undergrowth common in our eastern deciduous forests today is certainly not representative of pre-Columbian conditions.[45]

By repeatedly burning the vegetation, Native Americans also determined the composition of the forests as fire-resistant species, like oaks, were favored over fire-sensitive species, like maples. By the mid-1950s, all the original forests of New Jersey had been cut except for one small 65-acre woodlot. Recognized for its uniqueness as the sole remaining "virgin" forest in the region, it was purchased by Rutgers University and designated as the Hutcheson Memorial Forest. At the time, establishing this preserve was a minor media event. According to an advertisement in one national magazine, the Hutcheson Forest was a place where nature had been

> working for thousands of years to perfect this climax community in which trees, plants, animals, and all creatures of the forest have reached a harmonious balance with their environment. Left undisturbed, this stabilized society will continue to perpetuate itself century after century.[46]

Audubon, one of the leading environmental magazines, described the Hutcheson as

> a climax forest . . . a cross-section of nature in equilibrium in which the forest trees have developed over a long period of time. The present oaks and other hardwood trees [found there today] have succeeded other types of trees that went before them. Now these trees, after reaching old age, die and return their substance to the soil and help their replacements to sturdy growth and ripe old age in turn.[47]

As old trees in the Hutcheson Forest matured, managers assumed they would be replaced primarily by oaks—the same species that dominated the forest in the 1950s. After all, the Hutcheson Forest was supposed to be an old-growth, climax forest that would self-perpetuate ad infinitum, as long as modern man did not disturb it.

The Hutcheson Forest did not cooperate with the assumptions behind preserving it. In this undisturbed, supposedly climax forest, oaks did not grow up to replace the older oaks as those trees died of old age or disease. Instead, only maples grew to replace the original oaks. Additional research showed that the only way to maintain an oak forest as an oak forest was to burn it. The oaks, with their thicker insulating bark, are not affected by light ground fires set at the proper time of year—early in the spring or late in the fall—while the invading maples, with their thin bark, are easily killed. The supposed "forest primeval" was revealed to be, in fact, a product of aboriginal burning.[48]

The same is true in the western United States. Repeat photographs and stand-age analysis show that there was little old-growth forest prior to the elimination of native burning and active fire suppression. Moreover, the little old-growth forest that existed around 1850 was entirely different structurally and ecologically from what exists today. In the past, a few large, widely spaced trees were surrounded by a lush understory of grasses and flowering plants called *forbs*. Arizona's ponderosa pine forests, for instance, had just twenty to sixty trees per acre prior to European settlement, while today 1,000 trees per acre is common.[49]

Historically, most western forests would not support high-intensity, stand-replacing crown fires because frequent ground fires, set by native peoples, kept the forests open and park-like. Now that our forests have both matured and thickened, large-scale crown fires are becoming the rule, something that never happened before. Fire certainly structured most North American forests, but not lightning-caused holocaust-type infernos. Moreover, because we changed fire regimes, our forest ecosystems today are nothing like they were in the past, and their ecological integrity has been compromised.

This is true even in the coastal forests of the Pacific Northwest. In fact, all plans to save the endangered northern spotted owl are, to one degree or another, based on the assumption that the entire region was blanketed with old-growth forest (defined as trees more than 200 years old) before Europeans arrived.[50] But historical photographs and old stand maps show that in 1840, when large numbers of Europeans first began to occupy the Pacific Northwest, only 20 to 40 percent of the area supported old-growth forests. It may be hard to believe, but there is more old-growth forest today, despite a century of logging, than there was in 1800.[51] The reason is that for thousands of years native peoples structured all the Northwest's plant communities by repeatedly firing the vegetation.[52] The burning was so persistent that it created grasslands and open valleys in what would otherwise have been forested environments.[53]

The forests in the western United States are *not* self-perpetuating, climax forests; instead, most are born of fire. Even many of the forests in the Pacific Northwest need fire to regenerate.[54] Douglas fir, which presently dominates huge tracts of old-growth forests, will not regenerate in its own shade. That is to say, new Douglas fir trees are physically incapable of growing under a mature Douglas fir. The only way to maintain coastal Douglas fir forests is to burn them so that Douglas fir can then grow on the burned sites—in these forests, stand-replacing crown fires occurred at infrequent intervals.[55]

Conclusions

In this chapter we have disputed myths of nature undisturbed, the balance of nature, untouched wilderness, vast numbers of wildlife at Columbian contact, ecosystems, and old-growth forests. Our purpose is to suggest that those wishing to promote environmental protection, however that is defined, might want to jettison those myths and base their claims on less politicized versions of ecology.

One person who has done just that is Emma Marris in her book with the provocative title, *Rambunctious Garden: Saving Nature in a Post-Wild World.* The description of her book on its dust jacket describes her purposes as well as our own views of how to move beyond today's political ecology:

> A paradigm shift is roiling the environmental world. For decades people have unquestioningly accepted the idea that our goal is to preserve nature in its pristine, pre-human state. But many scientists have come to see this as an outdated dream that thwarts bold new plans to save the environment and prevents us from having a fuller relationship with nature. Humans have changed the landscapes they inhabit since prehistory, and climate change means even the remotest places now bear the fingerprints of humanity. Emma Marris argues convincingly that it is time to look forward and create the "rambunctious garden," a hybrid of wild nature and human management.
>
> In this optimistic book, readers meet leading scientists and environmentalists and visit imaginary Edens, designer ecosystems, and Pleistocene parks. Marris describes innovative conservation approaches, including rewilding, assisted migration, and the embrace of so-called novel ecosystems.
>
> *Rambunctious Garden* is short on gloom and long on interesting theories and fascinating narratives, all of which bring home the idea

that we must give up our romantic notions of pristine wilderness and replace them with the concept of a global, half-wild rambunctious garden planet, tended by us.[56]

Starting from the position Marris lays out rather than starting with myth and dogma allows for the possibility of creating policies that might actually work. But creating such policies will have to be done with a clear understanding of how political entrepreneurship works, a topic we take up in the next chapter.

3

Environmental Political Entrepreneurship

WE'VE SAT THROUGH dozens, possibly hundreds, of public meetings on environmental issues. At nearly every turn in the environmental process, there is opportunity to comment on proposed actions that may impact our lives in some way. Despite what many commenters think, this process is not a vote for a particular environmental outcome; it is a chance to make substantive comments and share information that may not be captured in the accompanying environmental analysis. Regardless, every group feels the need to show up in numbers to demonstrate their preferences: all-terrain vehicle rider clubs show up in boots and helmets, fishermen show up in waders, and Sierra Clubbers show up in Chaco or Birkenstock sandals. They were all alerted to the meeting by their respective organizations touting the importance of representing their cause. Vocal opponents or proponents will regale agency personnel with speculation about what will happen if the action is taken. In nearly every meeting we have attended, truths have been distorted, falsehoods have been spread, and facts have been exaggerated. Each group is simply using politics to look out for its own interests and to decrease the probability of negative impacts to the particular resource they care about.

Politics is about the relationships between individuals and groups in the political environment. We begin by assuming that individuals are utility maximizers who seek benefits from the political system. Politicians seek votes and re-election. Bureaucrats seek greater job security and bigger budgets. Interest groups and voters seek to have their wants implemented by the political system. Each political actor seeks something controlled by others in the system. For example, voters and interest groups want services from politicians and bureaucrats. Bureaucrats want greater revenues or budgets from politicians and taxpayers. Politicians want votes and other forms of support from citizens and interest groups. Democracy is not the process portrayed in the famous Norman Rockwell painting of a New England town meeting, which rather romantically depicts a humble citizen in the middle of an impassioned speech at a town meeting, as others from his community look on captivated. Instead, John E. Mueller explains the political process this way:

> As it happens, that misty-eyed, idealized snapshot has almost nothing to do with democracy in actual practice. Democracy is not a process in which one shining idea conquers all as erstwhile contenders fall into blissful consensus. Rather, it is an extremely disorderly muddle in which clashing ideas and interests (all of them "special") do unkempt and unequal, if peaceful, battle and in which ideas are often reduced to slogans, data to distorted fragments, evidence to gestures, and arguments to poses. Speculation is rampant, caricature is routine, and posturing is de rigueur; if one idea wins out, it is likely to be severely compromised in the process, and no one goes away entirely reconciled or happy. And there is rarely a sense of completion, finality or permanence: in a democracy as Todd Lindberg points out, "the fat lady never sings." It's a mess, and the only saving grace is that other methods for reaching decisions are even worse.[1]

Those who are expected to manage the results of lobbying and legislative battles are agency personnel—bureaucrats. Agencies or bureaucracies are nonprofit organizations whose major choices are shaped by political criteria such as budget allocations, internal politics, and responses to the agency's clientele. Because public agencies are creatures of the polity in general and politicians in particular, they are political agencies, not the omni-competent, impartial organizations envisioned by the designers of the civil service system.

Because political processes determine budgets, every bureau must choose fiscal strategies of survival and growth that make political—if not economic—sense. The successful political entrepreneur/bureaucrat must take actions that improve the agency's position vis-à-vis the legislative committee and the citizen-clients of the agency. Bureaucrats shape the demand for their own services from politicians and client groups. The preferred strategy of the bureaucrat includes many individual tactics. First, bureaucrats insulate themselves and their agencies from legislative scrutiny and control by obtaining earmarked funds. Second, bureaucrats add new services and supportive activities to those they already offer. Third, as nearly everyone knows, bureaucrats must spend all of their annual appropriations. Next, bureaucrats implement plans and programs that lead to elaborate procedures, rigidity, and inertia, but at the same time, bureaucrats endeavor to convince the public and legislative committees that the demand for their services is greater than anticipated and that their costs are higher than predicted—supplemental funds are usually requested as essential. And finally (and importantly), bureaucrats do everything in their power to avoid embarrassing errors that might reflect poorly on agency heads and their own missions.

Emotion over Science

Among the prominent pieces of legislation enacted shortly after the inaugural Earth Day, the Endangered Species Act of 1973 has created a variety of unintended, detrimental effects for property owners and endangered species alike. Controversy surrounding the act is less about the goal of protecting species on the verge of extinction—a widely appreciated end—and more about a perverse incentive structure that punishes property owners when certain species set up camp in their backyards. As a result of the law's incentive structure, occurrences of preemptive habitat destruction and the practice of "shoot, shovel, and shut up" have increased as land owners increasingly attempt to avoid the regulatory burdens and sanctions imposed by the law.[2] Since the law won't protect them, the ranchers sometimes protect themselves and their livelihood by shooting the wolf on their property—defeating the purpose of the law—and then quickly and quietly moving on. Notwithstanding deficiencies in the Endangered Species Act, groups like the Center for Biological Diversity rely on the law to force federal agencies to list more and more species. Once listed, endangered species are rarely removed, and the recovery rates and statistics of certain species and the ranges of rehabilitation are often framed to favor other environmental agendas rather than actual species protection.

The Endangered Species Act and the process of listing and delisting species have created a species that is anything but endangered: professional environmental litigants. The act has opened the door to groups like the Center for Biological Diversity to file lawsuit after lawsuit, knowing federal agencies won't have the time or resources to do anything but settle. The details of the settlements are hidden behind legal seals, which makes it impossible for the general public and even congressional investigators to discover them.

Science is sometimes a secondary consideration to groups who view species protection as a moral imperative. Kieran Suckling, the executive director of the Center for Biological Diversity, acknowledged that a scientific background is not necessary for activists hired by his organization. He writes,

> I think the professionalization of the environmental movement has injured it greatly. These kids get degrees in environmental conservation and wildlife management and come looking for jobs in the environmental movement. They've bought into resource management values and multiple use by the time they graduate. I'm more interested in hiring philosophers, linguists and poets. The core talent of a successful environmental activist is not science and law. It's campaigning instinct.[3]

With ever-increasing funds, an in-house legal team, and a "campaigning instinct," groups like the Center for Biological Diversity have inundated agencies charged with protecting species under the Endangered Species Act, their legal fees diverting resources intended for species protection and habitat restoration. They do it with great emotional intensity and fervor.

Catastrophe Sells

The term *environmentalist* did not emerge until the early 1960s.[4] About this time, industrial and agricultural production had reached an all-time high, bringing greater demand for energy, increased oil imports, greater consumption of natural resources, and more public awareness of industrial pollution.[5] Fueled in part by the counterculture movement that contributed to the civil rights movement and anti-Vietnam protests, modern environmentalism was "galvanized into an organized force" in the 1960s and '70s when popular television greatly expanded awareness of the environmental crisis and pioneering environmentalists like Rachel Carson forecasted global catastrophe.[6]

Environmental activists had highly visible disasters to use in the fight against industrial activity. Smog, river pollution, overconsumption, starvation, and health concerns stemming from environmental degradation created noteworthy stories that the media could highlight. The Cuyahoga River fire of 1969, for example, provided environmentalists with ammunition to pursue regulation of industrial waste. Before the 1969 fire, the Cuyahoga River in Cleveland, Ohio, was known as one of the most polluted in the United States, and it had caught fire thirteen times, beginning in 1868.[7] Unlike earlier fires on the river, the 1969 fire did not fatally injure anyone, and it resulted in less damage; however, the incident in 1969 received the most coverage due, in part, to "the developing precedence that sanitation held over industrial actions."[8] Cleaning up waterways became a priority of the environmental movement, later addressed by the Clean Water Act.

Other environmental disasters added momentum to the environmental movement and subsequent federal regulations. On January 28, 1969, a "blowout" on a Union Oil platform off the coast of Santa Barbara released more than 3 million gallons of crude oil into the Pacific; images of oil-topped waves and birds coated in black abounded in the national press.[9] In 1978 residents living around Love Canal near Niagara Falls, New York, were evacuated when remnants of an estimated 21,000 tons of buried industrial waste began to "bubble up into backyards and cellars."[10] Arent Schuyler, lecturer emeritus in environmental studies at the University of California, Santa Barbara, said, "People could see very vividly that their communities could bear the brunt of industrial accidents. They began forming environmental groups to protect their communities and started fighting for legislation to protect the environment."[11] Perceived disasters like the Santa Barbara oil spill and the 1969 Cuyahoga River fire are sometimes credited with inspiring Congress to pass the National Environmental Policy Act, created to manage environmental risks and disclose the impacts of industrial activity. As environmental leaders became more skilled in clearly, charismatically packaging their message, the news media was quick to snatch up the drama, nationalizing local environmental issues, thereby reaching audiences across the country.[12]

As claims about environmental disasters surfaced in nationwide broadcasts, a variety of environmentalists burst onto the scene with an assortment of doomsday scenarios. Foremost among the fear-casters was Rachel Carson. Using alarmist rhetoric, Carson targeted dichlorodiphenyltrichloroethane (DDT) as a primary threat to biodiversity and human health. Rachel Carson released *Silent Spring* in 1962, condemning DDT and modern organic pesticides as a threat to wildlife and humans alike. Carson sought to reform social attitudes toward nature that failed to prioritize environmental problems as she spoke of "man's assaults upon the environment" through air, water, and soil pollution.[13] DDT and similar compounds were of particular concern to Carson because they tended to persist in fatty tissue of animals after being ingested. "For the first time in the history of the world," Carson wrote, "every human being is now subjected to contact with dangerous chemicals, from the moment of conception until death."[14] "Unless we do bring these chemicals under better control we are certainly headed for disaster," Carson said. "The balance of nature is built of a series of interrelationships between living things, and between living things and their environment. You can't just step in with some brute force and change one thing without changing many others."[15] Rachel Carson's influence eventually contributed to the EPA's banning of DDT on January 1, 1973.[16]

Paul Müller, a Swiss chemist, was awarded the Nobel Prize in Physiology or Medicine in 1948 for discovering the insecticidal properties of DDT, which was first synthesized by an Austrian student in 1873.[17] During World War II, DDT spared thousands from a possible outbreak of typhus, which had claimed many lives during the Thirty Years War, destroyed remnants of Napoleon's Grand Army, and killed many during World War I.[18] In 1965 the National Academy of Sciences estimated that DDT prevented about "500 million [human] deaths that would otherwise have been inevitable."[19] According to the World Health Organization, DDT "killed more insects and saved more people than any other substance."[20]

Despite DDT's ongoing potential to limit malarial outbreaks in developing countries, many environmentalists continue to insist on following through on Rachel Carson's legacy of banning the chemical. As a result of such efforts, the United Nations has considered global bans on DDT. In 2001, after withdrawing from the Kyoto Protocol, President George W. Bush signed the Global Convention on Persistent Organic Pollutants, a UN attempt to stop member states from producing and exporting certain chemicals like DDT.[21] As an exception, DDT would be allowed for malarial control, but countries would have to show that the chemical was being phased out.[22] An absolute ban on DDT would disproportionately affect developing nations, where malarial outbreaks are more common. Some suggest that sacrificing human lives by withholding DDT helps reduce the risks of overpopulation and more environmental harm. Former organizer for Earth First! Jeff Hoffman, for instance, wrote the following in a blog post, "Malaria was actually a natural population control, and DDT has caused a massive population explosion in some places where it has eradicated malaria. More fundamentally, why should humans get priority over other forms of life? . . . I don't see any respect for mosquitos [sic] in these posts."[23] Such remarks, while not necessarily representative of mainstream environmentalism, highlight a pervasive, ideological aversion to human impacts on nature.

Ideological roots behind Hoffman's statement and general aversion to human influence on the planet can be traced to Paul Ehrlich's *The Population Bomb,* first published in 1968.[24] Ehrlich, a professor of entomology at Stanford University, was another outspoken fear-caster during the environmental movement's nascent stages. As a high school student in New Jersey, Ehrlich believed that the butterflies he was interested in studying were disappearing and blamed increasing population, the building of more and more subdivisions, and the spraying of DDT.[25] His concerns were compounded after traveling to India. Ehrlich recounted his life-changing trip, writing,

> The streets seemed alive with people. People eating, people washing, people sleeping. People visiting, arguing, screaming. People thrusting their hands through the taxi window, begging. People defecating and urinating. People clinging to buses. People herding animals. People. People. People. People.[26]

In response to his experience with the slums in India, Ehrlich "proposed limits on population, including economic sanctions and forced sterilization."[27] Ehrlich feared not only the effects of starvation, but also the increased use of fertilizers and pesticides, the clearing of forests, and the depletion of resources to accommodate a burgeoning global population.[28] Ehrlich feared a global environmental crisis and even put a timeline on his catastrophic predictions: "Sometime in the next 15 years the end will come and by the end I mean an utter breakdown of the capacity of the planet to support humanity."[29] Environmentalists were divided regarding their support for what appeared to be a modern form of eugenics and scientific racism, with some deciding to focus on Ehrlich's views of overconsumption and others supporting Ehrlich as one speaking unpleasant truths.[30]

In a less misanthropic manner, Congressman Stewart L. Udall had warned of the dangers of overconsumption of natural resources in *The Quiet Crisis,* published in 1963.[31] Udall worried that the lifestyle of abundance and economic prosperity would lead to a gradual loss of natural habitat. Any sense of well-being after World War II, in Udall's view, was deceptive, causing society to ignore a "steadily degrading natural environment."[32] Udall's influence went well beyond activism and played a significant role in governmental policy decisions. When his book was published, Udall was serving as Secretary of the Interior (he served under the Kennedy and Johnson administrations). In this role, Udall helped pass the Wild and Scenic Rivers Act and created the Land and Water Conservation Fund.[33]

Changing the Arena from States to the Federal Government

A useful strategy for reducing lobbying costs is to elevate an issue to the national level, rather than having to fight separately in the fifty states. The environmental movement and the federal legislation enacted in the 1960s and 1970s are usually credited with resolving problems associated with air pollution, water quality, and industrial waste. While it may be true that these regulations contributed to some immediate environmental gains, state and local regulations were already in place that had contributed to declining pollution levels. Discussing what he calls the "fable of federal environmental regulation," Jonathan Adler explains a common assumption, which holds that a lack of action on environmental issues by the states necessitated federal intervention.[34] While many perpetuated this fable using pictures of the Cuyahoga River burning and similar environmental disasters as evidence for deteriorating environmental quality, history was telling a different story.

Industrial waste in rivers was a common problem during the early 1900s.[35] Industrial river pollution was seen not as a cause for concern, but rather as a "sign of progress" in the early industrial movement in the United States.[36] The Cuyahoga River in Ohio was no exception. Like the Cuyahoga, many rivers near industrial centers had burned prior to the infamous 1969 blaze.

The 1969 event garnered more attention because of heightened awareness of environmental issues and an increase in U.S. wealth that resulted in a greater demand for environmental quality.[37] This greater demand, however, did not mean that states were altogether neglecting to clean up their rivers. Restorative efforts with the Cuyahoga River began before the *Time* article on the fire of 1969. In 1968, local voters approved a $100 million bond initiative to fund a river cleanup, and local efforts to improve conditions in the river had begun after a 1952 fire.[38] Moreover, many factories were shutting down or cutting back in the years leading up to the 1969 fire. While this information alone does not justify the levels of pollution in the Cuyahoga that caused $100,000 of estimated damages, it shows that the community was already engaged in environmental reform before the federal legislation was passed. Writing for Cleveland Historical, Michael Rotman summarizes this point: "The '69 fire . . . was not really the terrifying climax of decades of pollution, but rather the last gasp of an industrial river whose role was beginning to change."[39]

Nationwide, states enacted a variety of environmental policies before federal legislative floodgates opened in the 1960s and 1970s. Prior to the Clean Water Act and federal wetland regulation, "all fourteen states in the continental U.S. with more than ten percent of their land area in wetlands . . . had adopted wetland protection measures," and by 1966, every state had adopted some form of water pollution legislation.[40] Cincinnati and Chicago adopted some of the first air quality laws, with smoke control ordinances in 1881, and many other cities followed suit after World War II.[41] Prior to the Clean Air Act Amendments in 1970, recorded levels of key pollutants were declining, indicating that effective pollution control measures were already in place in many states.

Recognize and Take Advantage of Legislative Oversight and Control Loss

To Congress falls the task of monitoring the agencies. This monitoring task, known as legislative oversight, is assigned to committees and subcommittees, whose members sit on as many as a dozen other committees. Carrying out the monitoring function introduces more problems for rational policy-making. The first problem has to do with the members of Congress: They are busy.

Between 5,000 and 10,000 bills are typically introduced each congressional session. Thus, members of Congress are likely to pay attention only to those committees that hold a particular fascination or serve to promote the congressperson's agenda.

In a process of natural selection, members serve on committees that provide benefits to local interests. Representatives from districts with military bases and defense contractors sit on armed services committees, especially military appropriations committees. Representatives from farm states serve on agricultural committees, while members from urban areas sit on urban transit committees. The result is that a congressperson may easily develop a protectionist and even expansionist mentality for the bureaucracy he or she oversees.

Even if politicians overseeing agencies were somehow neutral and objective about the agency's mission, the information they receive from the agencies is likely to be misleading or slanted. Effective oversight requires information about the agency's actions and results, the best source of which is the agency itself, and the agency controls much of that information. Just as most people find it impossible to argue against their own best interests, most political managers find it impossible to present information about themselves in ways that might cause harm. Thus, information is likely to be selected that presents the agency in a good light, while negative information is kept out of sight.

The process by which bureaucrats dilute and slant information about their own agencies is called "control loss," and it comes from many sources. One of the best explanations of the process comes from economics professor David B. Johnson:

> Analysts can select data or methodologies that will produce results known to be favored by the department head. Studies that give the "wrong" answers are ignored, while those that give the "right" results are "sent forward." Control loss can occur when bureaucrats summarize voluminous studies into a few pages of "executive summaries" that become smaller as a report makes its way up the bureaucratic ladder. The most damaging data, comments, and qualifying phrases can be ignored, while the positive comments are emphasized . . . One of the most serious sources of control loss is the difficulty of imparting detailed instructions to bureaucrats. Bureaucracies exist because it is

difficult to define their services sufficiently well so that they can be contracted out to private firms. This, of course, means that it is also difficult to give precise instructions to administrators in the bureaucracies. Bureaucrats are able to give their own interpretations to the general instructions they receive and, thus, gain valuable flexibility.[42]

Management plans developed by the U.S. Forest Service (USFS) provide an excellent example of the kinds of control loss that Johnson describes. The plans provide many opportunities to select, ignore, or slant data and to hide information damaging to a supervisor's goals. Reviews of agency policies provide more opportunities for control loss. At Yellowstone National Park, for example, the person chosen by the U.S. National Biological Survey to prepare a report on whether the Yellowstone elk herd was damaging the northern range in Yellowstone was a U.S. National Park Service employee who had written extensively for the Park Service justifying the park's elk policy. Not surprisingly, he reported that no damage was occurring.[43]

Another reason for control loss is what might be called the "management information problem." Acquiring the information needed for environmental protection sounds simple, but it is monumentally difficult for several reasons, including political pressures, poor data, contradictory scientific theories, and planning horizons with schedules that keep data out of the decision-making process.

Political Pressure

Environmental issues, like all political issues, arise through a developmental process. First, a new concern appears. This concern is debated, and political action is taken. Then, advocates who were organized to create pressures to "do something" remain active to preserve their causes and the political programs that address them. As the politics shift from the original issue to government programs, attention shifts from the original issue to the programs. This is the normal process of interest group politics, and environmental politics are not immune.

Political management raises a number of election-related problems because politicians must face election from time to time, and politicians appoint agency heads. This political requirement makes administrators more attentive to those most likely to influence the next election—organized groups, not individuals. In addition, political appointees pay attention to what pacifies the electorate rather than what is good policy. It matters little whether the policies are based on sound science or on environmental myths. Instead, there are strong incentives to invent policies that sell, sometimes regardless of effects.

Simmons, one of the authors, witnessed political pressures while working in the Interior Department's Office of Policy Analysis in the early 1980s. At one point, the assistant secretary for land and water, a former university professor with a Ph.D. in economics, argued for continuing water subsidies for farmers in the western United States. When he was asked how, as an economist, he could justify those subsidies he replied, "As a politician I have to."

A popular theory of politics is that competing groups check the excesses of each other, and the general good emerges. This competition may exist sometimes, but at other times, groups cooperate instead and exchange support instead of competing. Thus, benefits to one group are matched by benefits for another. Off-road vehicle enthusiasts are able to use public lands not because theirs is the highest-valued use of the land (although it may be) but because they can garner the support of cattle grazers, dam builders, and the tourism industry—all groups who are organized to influence the political allocation of the public lands. In exchange, off-road vehicle groups support the wants of the other groups. The result is protection of the wants and special privileges of organized groups—not necessarily protection for species, watersheds, or the land.

Take Advantage of Planning Horizons

The structure of the planning and management process of federal and state agencies creates opportunities for alert political entrepreneurs, often those within the bureaucracy. Forest and grazing plans, environmental impact statements, and court decisions often restrict a manager's abilities to make important decisions "on the ground." More than thirty years ago, Dick Behan, former dean of forestry at Northern Arizona State University, argued that central planning of national forests is impossible and often disastrous because it does not incorporate time- and site-specific information.[44] His claims are more valid today than they were at the time, and Forest Service planning is stuck in conflicting data, opposing perspectives, and lawsuits. All government agencies assigned with managing the environment face similar difficulties.

Employees of the BLM face serious practical problems as well. They need information about range conditions. Which areas are prone to erosion or easily compacted? Do some areas recover more quickly from intensive grazing than others? Do these evaluations differ depending on the time of year and climatic conditions? To manage effectively, this information must be acted upon as particular conditions of time and place change. Bureaucratic management, however, does not allow for time- or site-specific adjustments or for site-specific information. Instead, stocking rates and grazing dates are established centrally without regard for site-specific information.

Despite the difficulty of collecting and using information, as well as the possibility that laws mandating action might be merely policy symbolism, law and court orders require the BLM to formulate extensive management plans for each grazing district. Developing these plans requires intensive efforts involving teams of planners who are expected to strike an acceptable balance between use and preservation in the face of competing political constituencies. As conditions change, the information guiding the management plan becomes outdated, but policies based on the plan continue in force until new information and plans can be developed. Such processes take years, sometimes decades, and by the time the plans are revised, a new future requiring new information has presented itself. Therefore, the hand of the past continues to guide the new and different present. In such situations, present policies produce good management only by serendipity.

Recognize the Potential Costs of Risky Choices

The decisions made by bureaucrats are often critical to the health of economies and ecosystems. Unfortunately, bureaucrats receive few direct rewards for wise choices, yet they are soundly criticized when their work conflicts with standard procedures or beliefs. Arduous debates precede any large-scale innovation and even small ones. Perceived failures can lead to the dismissal and replacement of agency personnel. Innovation is a chancy business, and the reasons for a specific experiment failing are far more evident at the end than at the start of a project. The result is far more opportunities for criticism than praise. Thus, the rational bureaucrat finds it safer to avoid risk and stick with what is safe. Scientists funded by government agencies will also tend to choose research subjects that will not get them in trouble with their funders.

In a seminar Simmons attended at Utah State University (USU), a former dean of the USU College of Natural Resources related a telling example of the danger of taking risks. He told of an entrepreneurial BLM manager in Utah who devised a new, efficient, and accurate means for estimating range quality. The manager had been told in his range science classes at USU that the BLM's required method was completely unreliable—different people reported different levels of quality for the same range using the method. The method, called *ocular reconnaissance,* was little more than looking around and making estimates without any objective measurements. The manager's new system, however, was nearly as quick while providing accurate, useful information, and it earned him the "Range Scientist of the Year" award from the Utah Range Science Society. Nevertheless, the BLM transferred him because he refused to use the old, useless method. He was told the agency collected the information only because it was under a court order, not because it affected management choices. It may be useful to remember that not all private entrepreneurs succeed, so we should not expect all public ones to succeed either.

Another example we know of involves Richard Keigley, an entrepreneurial plant ecologist with the Park Service who studied cottonwoods along the Lamar River in northeastern Yellowstone. By studying core samples from the cottonwoods, he concluded that the trees had grown normally only during times in the past when the park's managers had reduced the size of its elk herd. The problem for park management was that his conclusions directly contradicted Yellowstone's natural regulation management plan.[45]

The *Los Angeles Times* reported that the Park Service attempted to transfer Keigley from Yellowstone and refused to allow him to publish his data.[46] In testimony before the U.S. House of Representatives Resources Subcommittee on Parks and Public Lands, Keigley claimed that even after his position was moved to the NBS and later to the U.S. Geological Survey's Natural Resources Division, the Park Service continued to prevent him from doing research anywhere in the Yellowstone ecosystem.[47] Columnist Alston Chase reported that Park Service personnel refused to issue Keigley a permit to conduct research in the park and used political pressures within the Clinton administration to prevent him from studying areas outside the park, even though the areas were not under Park Service or Interior Department control.[48] Spokespersons from the Park Service claimed that Keigley's research plan was poorly conceived and not defensible, even though it had undergone a rigorous review process outside the Park Service and had drawn praise for its originality and thoroughness.[49]

Conclusion: Bad People or Good Politics?

Strategic use of political processes is neither new nor confined to environmental politics. Entrepreneurs in any political arena employ strategy and tactics to achieve their ends. We suggest that the careful analyst should distinguish between announced ends and those actually pursued. Thus, a Clean Air Act might be more about protecting coal interests than cleaning the air; many environmental laws are used to rally public support and generate funding rather than to accomplish environmental goals. Myths and symbols may substitute for science if they benefit a political agenda.

Strategic political actors produced all the major environmental legislation and regulation in the United States. Successful environmental policies and laws have been enacted, but they occur less frequently than they could, and they come at extravagant cost. That claim suggests that there might be room for political entrepreneurship that generates better outcomes.

4

The Politics of Nature

WILDERNESS IS A modern cultural and political invention. It is a designation based on arbitrary parameters of naturalness and the absence of humans. For some, wilderness exists only in the mind's eye, so elusive you really know it only when you see it. Historian William Cronon explained it this way: "Go back 250 years in American and European history, and you do not find nearly so many people wandering around remote corners of the planet looking for what today we would call 'the wilderness experience.'"[1] America's early settlers and pioneers often considered themselves in a fight with nature for survival, prosperity, and safety. Taming the wild was viewed as progress. Only after people became relatively wealthy did wild nature seem like an ideal worth pursuing.[2] The historian Roderick Nash summarized early attitudes of American settlers toward nature: "The pioneer, in short, lived too close to wilderness for appreciation."[3]

Absence Makes the Heart Grow Fonder

As the United States urbanized and industrialized, Americans came into contact with nature less frequently and in fewer numbers.[4] Advancements in technology and decreased contact with nature eventually helped create what Cronon called a "wilderness cult," or a movement that resembled eighteenth-century primitivism—"the belief that the best antidote to the ills of an overly refined and civilized modern world was a return to a simpler, more primitive living."[5]

Cronon identified a tendency among romantic writers to depict nature as "a symbol of God's presence on earth."[6] Mountains, peaks, and hills were converted into cathedrals because of the awe-inspiring emotions such landscapes evoked. John Muir, the founder of the Sierra Club, helped guide an evolving wilderness movement that adopted a religious stance toward nature. Notwithstanding his spiritual views of nature, Muir did not oppose people building houses, raising crops, and conducting low-key mining operations in such areas, so long as they were not "mere destroyers . . . tree-killers, wool and mutton men, spreading death and confusion."[7] However, Muir's enduring legacy lies not in his willingness to accommodate low-impact forms of living in nature but in his views of wilderness as sacred space.

Others in the wilderness movement feared not only that resource exploitation, industrialization, and urbanization of frontier landscapes threatened the spiritual values with which early romantics imbued nature, but that such trends also threatened an unsettled frontier whose passing would carry away a piece of American identity. In 1893, witnessing an expanding, industrialized population, Harvard historian Frederick Jackson Turner declared the American frontier "closed." To Turner, nature and the uninhabited frontier embodied qualities of American character such as independence, self-reliance, and democracy itself.[8] Turner believed wild lands and frontier landscapes had shaped not just the American people but their institutions as well. With the closing of the frontier, he implied there would be a difficult transition to something other than the America that had developed as a result of exploration and settlement of the frontier.

Perhaps in response to a perception of shrinking western landscapes and the beginning of the transition Turner expected, more Americans began "going camping, starting Boy Scout troops, and reading Jack London stories about hardscrabble life in Alaska."[9] Theodore Roosevelt created his own brand of rugged individualism, promoting survival in the wilderness, hunting, fishing, and testing oneself against the wild. Roosevelt sought to establish responsible use of resources to ensure prosperity for future generations, which was a view quite distinct from Muir's preservation ideology. During a governors conference convened by Roosevelt, he framed his position on environmental protection in these words: "We look upon these resources as a heritage to be made use of in establishing and promoting the comfort, prosperity, and happiness of the American people, but not to be wasted, deteriorated, or needlessly destroyed."[10] Roosevelt used his presidential power to designate "150 National

Forests, the first 51 Federal Bird Reservations, five National Parks, the first 18 National Monuments, the first 4 National Game Preserves, and the first 21 Reclamation Projects."[11]

Wilderness in Our Own Image

Another response to perceptions of a vanishing frontier was establishing national parks. Muir thought of himself as a John the Baptist figure for nature and used spiritual conceptions of certain lands to justify the idea of parks devoid of human inhabitants. To Muir, preserving nature meant preserving expressions of the divine, or "sparks of the Divine soul."[12] Indigenous tribes "seemed to have no right place in the landscape," in Muir's mind.[13] Forcefully evicting indigenous Miwok tribes from Yosemite with armed militias helped Muir create his ideal of an uninhabited sanctuary in the wilderness, even though this ideal was completely fabricated.[14] For generations, Native Americans had cultivated, burned, hunted, and lived in areas that became national parks, even under protection of treaties with the U.S. government. In violation of treaties, and in an effort to leave nature "undisturbed," expelling natives became a routine practice in the development of national parks. Muir viewed removing Native Americans as a worthwhile means to promote his aesthetic and spiritual vision concerning the land.[15]

Early conservationists typically had significant wealth and prominence in society, luxuries that enabled them to pursue forms of recreation in nature that were alien to the experience of the working class. "Country people generally [knew] far too much about working the land to regard unworked land as their ideal," writes Cronon. "In contrast, elite urban tourists and wealthy sportsmen projected their leisure-time frontier fantasies onto the American landscape and so created wilderness in their own image."[16] Virtually all the members of prominent organizations such as the Sierra Club (organized in 1892) and chapters of the National Audubon Society in New York and Massachusetts (organized in 1896) were white Anglo-Saxons, "well bred hunters, fishermen, and campers." Wealthy nature enthusiasts also helped establish Roosevelt's Boone and Crockett Club in 1887, an organization dedicated to protecting game animals and their habitat for hunting.[17]

Initial conservation efforts geared toward recreational activities for the wealthy also affected marginalized groups. Restrictions on land use and hunting practices limited the resources that immigrants, racial minorities, the destitute, and indigenous tribes relied on for sustenance. The director of the New York Zoological Park, William Temple Hornaday, referred to Italians, southern Negroes, poor whites, pot hunters, gunners, and indeed "every man who still shoots game" as soldiers in a "regular army of destruction."[18] In an article discussing what they call "environmentalism's elitist tinge," environmental historians Joseph E. Taylor III and Matthew Klingle describe how, when it came to preserving tracts of land for specific recreational experiences, wealthy hunters and fishers had "the money and political connections to persuade state and provincial legislatures to do their bidding."[19] Conservation became a tool that restricted poor peoples' access to nature as "states mandated the purchase of licenses, prohibited equipment favored by people who depended on wildlife for subsistence, and restricted the spaces and times of harvest."[20]

Clashing Land Ethics

Gifford Pinchot became another powerful political force when he entered the conservation movement as the first chief of the U.S. Forest Service (Forest Service) in 1905. Unlike Muir, who believed that wild lands should be preserved for their innate value regardless of their usefulness for society, Pinchot tempered Muir's and even Roosevelt's romanticism by advocating conserving the nation's forest reserves (later called National Forests) for producing "whatever it can yield for man." He promoted a utilitarian vision of the forests under which forests were to be managed for continuous cropping, which is why the U.S. Forest Service is a part of the U.S. Department of Agriculture.

As Teddy Roosevelt's Forest Service director, Pinchot used the term *wise-use* to describe his policy of resource extraction and land management. Pinchot said, "The object of our forest policy is not to preserve the forests because they are beautiful . . . or because they are refuges for the wild creatures of the wilderness . . . but the making of prosperous homes. Every other consideration comes as secondary."[21]

Before long, Pinchot's wise-use ethic bumped up against Muir's religious devotion to preservationism. The damming of the Tuolumne River, inside

Muir's personal sanctuary, the Hetch Hetchy Valley in what is now known as Yosemite National Park, represents one of the earliest national environmental controversies in the United States. The city of San Francisco sought to dam the river to accommodate an increased water demand. When San Francisco Mayor James D. Phelan applied to designate the valley as a reservoir site, Interior Secretary Ethan A. Hitchcock initially refused to "violate the sanctity of a national park."[22] Vehemently opposing the construction of the dam, Muir sparked a national protest campaign, which helped shift cultural attitudes toward nature.[23] "Few would have questioned the merits of 'reclaiming' a wasteland like this in order to put it to human use," writes Cronon. "Now the defenders of Hetch Hetchy attracted widespread national attention by portraying such an act not as improvement or progress but as desecration and vandalism."[24]

The struggle for the dam's approval highlighted divisions between two distinct land ethics. At the time, "Muir was considered the supporter of a shortsighted, elitist preservationist philosophy," whereas Pinchot was perceived as a "progressive conservationist whose views were in step with the prevailing public sentiment that natural resources should be used to enrich the lives of all Americans, not just the wealthy."[25] Statements from Pinchot and Muir regarding the dam show the differences in how they viewed nature. In 1913 Pinchot said, "As to my attitude regarding the proposed use of Hetch Hetchy by the city of San Francisco . . . I am fully persuaded that . . . the injury . . . by substituting a lake for the present swampy floor of the valley . . . is altogether unimportant compared with the benefits to be derived from its use as a reservoir."[26] Muir's words regarding the proponents of a dam are more ideological. He said, "These temple destroyers, devotees of ravaging commercialism, seem to have a perfect contempt for Nature, and instead of lifting their eyes to the God of the Mountains, lift them to the Almighty Dollar."[27] These comments planted the seeds for what would become a continuous struggle between competing narratives in the modern environmental movement: ideological environmentalism, or protecting nature for nature's sake, and utilitarian conservationism, or managing resources for human use.

After an earthquake and fire in 1906, the urgency of the city's request for additional fresh water, along with public sympathy, convinced U.S. Interior Secretary James R. Garfield to approve the project.[28] "Domestic use," wrote Garfield, "is the highest use to which water and available storage basins . . .

can be put."²⁹ President Woodrow Wilson signed the Raker Act in December 1913, which permitted the flooding of the Hetch Hetchy Valley by the O'Shaughnessy Dam.³⁰ Even in defeat, Muir felt as though he was "doing the Lord's battle" by opposing the dam.³¹ The fight over Hetch Hetchy Valley became "the battle cry of an emerging movement to preserve wilderness."³²

Another influential figure of the time was Aldo Leopold, an ecologist who worked under Gifford Pinchot for the Forest Service. He saw wilderness as "the stuff that America [was] made of" and feared the impacts of automobiles, roads, and consumerism on wild lands.³³ In an effort to preserve remote areas of land that seemed of little economic value, Leopold helped organized the Wilderness Society in 1935.³⁴ Originally trained in conservation, Leopold ultimately adopted a preservationist outlook following a predator-control hunt in 1909. After shooting a mother wolf, Leopold recounted walking up to the dying animal and seeing "the fierce green fire dying in her eyes."³⁵ In 1949, almost forty years after the event, Leopold's most significant contribution to the environmental movement came one year after his death when his work, *A Sand County Almanac,* which has been called a "[cornerstone] for modern conservation science, policy, and ethics," was published.³⁶ In the book, Leopold called for a new land ethic that would change "the role of homo sapiens from conqueror of the land-community to plain member and citizen of it." He added, "A thing is right when it tends to preserve the integrity and stability and beauty of a biotic community . . . [and] wrong when it tends otherwise."³⁷ Leopold insisted that large tracts of land be "deeded to the commons," an idea opposed by some contemporary environmentalists like Jay Darling, founder of the National Wildlife Federation.³⁸ Leopold hoped to preserve pristine areas for recreational use and enjoyment in the future. To this day, Leopold's vision, alongside Muir's and Roosevelt's, is used as a philosophical foundation for land management policies and environmental agendas.

Earth Icon "Changes the Rules of the Game"

Although Congress enacted significant legislation during the 1960s (the 1963 Clean Air Act, the 1965 Clean Water Act, and the 1964 Wilderness Act are the most prominent), the modern environmental movement may be traced to a decidedly non-natural event, the Apollo 8 space mission in 1968. During that mission, astronaut William Anders took a photograph of the Earth, producing what has been called "the most influential environmental photograph ever taken."[39] Earthrise, an icon of the modern environmental movement, gave viewers a conception of the Earth as a small object in a vast sea of space.[40] To many in the environmental movement, "Earth in its surrounding dark emptiness not only seemed infinitely beautiful, it seemed infinitely fragile."[41] This iconic image of a fragile planet with limited resources resonated among those within the environmental community who sensed an impending doom because of mankind's encroachments against a perceived balance in nature. In the 1960s and 1970s, when the modern environmental movement was building up a head of steam, alarmists tried to describe the Earth as fragile, delicately balanced, and on the verge of destruction—all thanks to humans. To many environmentalists, "the rapid pace of modernity," using historian James Turner's words, "threatened to tear the fabric of life apart in a matter of years."[42]

Efforts within the environmental movement to save a fragile Earth from perceived pollution and overconsumption resulted in the celebration of Earth Day on April 22, 1970, one of the most significant public events in the early years of the environmental era. The counterculture of the 1960s, with its antiwar and civil rights movements, laid a foundation for protests against consumerism, waste, corporate irresponsibility, and ecological exploitation.[43] Dennis Hayes, one of the primary organizers of the first Earth Day, described the purpose of the event. He said, "What we were trying to do was to create a brand new public consciousness that would cause the rules of the game to change."[44]

The rules of the game began to change with the magna carta of environmental legislation: the National Environmental Policy Act (NEPA) of 1969.[45] Signed into law by President Richard Nixon, NEPA passed in the House of Representatives and the Senate with overwhelming support. Among its many effects, the law required federal agencies to assess and control potential environmental impacts during the decision-making processes for projects on federal lands. NEPA also required agencies to meet certain national environmental goals and policies. Historian James Turner summarizes the effects of NEPA:

> Before NEPA, the federal government had few obligations to consider the environmental ramifications of road-building, power plant siting, dam construction, or other such activities. The National Environmental

Policy Act changed that. Section 102 required a detailed environmental impact statement for any activity undertaken, permitted, or funded by the federal government that had significant environmental consequences. To meet NEPA's procedural requirements, the agency had to undertake an interdisciplinary analysis, document the potential environmental consequences, consider alternatives, undertake consultation with local and state governments and federal agencies, and engage the public in the process through hearings and comment periods. It was an unprecedented tool for citizens to monitor and affect the federal government's activities.[46]

The political and environmental effects of NEPA and the ways that political entrepreneurs use it are described in detail in Chapter 5. But we briefly note here that NEPA accomplishes less than it might at a huge cost. Its ineffectiveness, like that of other environmental legislation, can be traced to the dominant thought processes of the time period, which emphasized a balance of nature. Indeed, NEPA assumes a static baseline of environmental change and vastly oversimplifies complex ecological processes.

Shortly after NEPA was enacted, President Nixon reorganized the "three federal Departments, three Bureaus, three Administrations, two Councils, one Commission, one Service, and many diverse offices" into one cohesive organization, the Environmental Protection Agency (EPA).[47] Nixon's 1970 State of the Union Address prioritized a realignment of environmental protection:

> Restoring nature to its natural state is a cause beyond party and beyond factions. It has become a common cause of all the people of this country. It is a cause of particular concern to young Americans because they, more than we, will reap the grim consequences of our failure to act on programs, which are needed now if we are going to prevent disaster later.[48]

In his statement when he signed the bill authorizing NEPA, Nixon said, "the 1970s must absolutely be the years when America pays its debt to the past by reclaiming the purity of its air, its waters, and our living environment."[49]

Over the next decade, twenty-three federal environmental acts were signed into law.[50] New laws enhanced and altered environmental protections established in the 1963 Clean Air Act, the 1965 Clean Water Act, and the 1964 Wilderness Act. Passed shortly after NEPA, the most prominent laws included a revised 1970 Clean Air Act to "set tighter standards for utility and auto emissions," a revised 1973 Clean Water Act to "bring a new measure of health to the nation's rivers, bays, and estuaries," and the Endangered Species Act (ESA) of 1973, which supposedly increased protection of endangered species and created a comprehensive system for listing species as threatened or endangered.

New rules for public land management became effective when the Federal Land Policy and Management Act (FLPMA) of 1976 gave the U.S. Bureau of Land Management (BLM) authority to manage public land for multiple uses (borrowing from the Multiple Use and Sustained Yield Act of 1960) and values.[51] Under the law, the BLM was required to keep an inventory of areas of ecological concern and conduct wilderness reviews, thereby establishing wilderness study areas.[52] The law repealed more than 2,000 old statutes for public lands management but left a host of other laws on the books, including the Mining Law of 1872, the Mineral Leasing Act of 1920, and the Taylor Grazing Act of 1934, which was an effort to fight overgrazing.[53] The law also created a process patterned after the environmental impact statements required by NEPA. The process required public input, interdisciplinary analysis, and intergovernmental consultation when planning uses for areas of public land. Land could be sold or developed for private interests under the law, but only after following NEPA procedures (e.g., conducting an environmental assessment, drafting an environmental impact statement (EIS), and jumping over other time-intensive, bureaucratic hurdles).

NEPA, FLPMA, and other environmental laws gave environmentalists standing in federal courts.[54] Environmental activists began to realize that they had been given a set of powerful tools. Manipulating the legislation itself proved to be more powerful than fighting the outcomes of the legislation. Negative outcomes, including increased costs to taxpayers, lost revenues, and increased governmental and private legal expenses, did not matter: Organizations were now able to deliver reliably on promises to constituencies.

Judicial appeals became an avenue through which environmental groups could circumvent the traditional route for reform (e.g., canvassing representatives for support, mobilizing constituents, and lobbying special interests). Following NEPA and nearly a decade of enhanced environmental regulation, environmental groups employed a variety of litigation tactics, developed and spearheaded by the Environmental Defense Fund, the National Resources Defense Council, and the Sierra Club Legal Defense Fund (EarthJustice). Thanks to increasing regulations, environmental groups were able to "challenge a range of government activities, including siting nuclear power plants, auctioning oil and gas leases, and permitting pesticide application programs."[55]

Following the rush of environmental legislation in the 1970s, environmental organizations increasingly relied on enhanced federal legislation, professional lobbying, and the courts to pursue objectives relating to, among other things, endangered species and wilderness designations. Legal victories and additional regulations helped organizations like the Wilderness Society establish themselves as mainstream contributors in the environmental movement. In addition to heavily staffed, nonprofit, corporate environmental organizations (e.g., Sierra Club, Natural Resources Defense Council, Environmental Defense Fund), sometimes referred to as "Gang Green," many smaller environmental schools of thought have emerged as grassroots contributors to modern environmentalism (e.g., deep ecology, ecofeminism, social ecology, the environmental justice movement, intergenerational equality).

As the national organizations of the environmental movement became increasingly centralized, organized, and oriented toward federal regulations and legal tactics, smaller environmental groups like Earth First! emerged. These groups preferred more radical means to protect the environment. They played an important strategic role by making the mainstream groups appear to be reasonable. Part of their strategy was to keep the mainstream groups from sliding away from their core environmental mission.

Deserting the Main Stream

Beginning in the 1970s, the environmental movement spawned biocentric ideologies that valued biodiversity and ecological integrity above conventional concerns for human health and well-being. Foremost among the biocentric ideologies was deep ecology, a nature ethic pioneered by Arne Naess. The term *deep ecology* sprang from Naess' belief that mainstream environmentalism "with its pragmatic and reform-oriented agenda" was shallow, only protecting human health and the interests of the privileged.[56] Naess' posited that all species have an inherent right to live regardless of their potential uses for humanity and deserve an equal footing with humanity in nature.

The deep ecology movement created a philosophical base for other biocentric, grassroots groups. Dissatisfied with the tactics of larger organizations (e.g., Wilderness Society, Sierra Club) in enacting wilderness protections, Dave Foreman founded Earth First! under the motto, "No compromise in defense of mother earth."[57] Historian James Turner credits deep ecology with influencing "Earth First's embrace of a biocentric world-view, its decentralized organizational structure and its strong critique of the national environmental groups."[58] Earth First! abandoned traditional policy tactics when pursuing environmental protection and sought controversy with illegal actions to bring attention to issues of biodiversity and wilderness protection. In 1985 Foreman published *Ecodefense: A Field Guide to Monkeywrenching*, which laid out strategies to spike trees (driving large spikes into trees to discourage cutting them), disable bulldozers, and sabotage roads (among other tactics) in an effort to save Earth from an "overconsumptive technological society."[59]

The tactics promoted by Earth First! and founder Dave Foreman had severe repercussions for at least one worker in the timber industry. While working for Louisiana Pacific paper in 1987, George Alexander was unfortunate enough to hit a spike in a section of a redwood log with his chainsaw and received severe face lacerations, cuts on both jugular veins, and loss of teeth from the accident.[60] Without taking credit for the spike, Foreman's response to the incident indicated that Earth First! viewed human harm as collateral damage in a more important struggle: "It is unfortunate this worker was injured and I wish him the best . . . But the real destruction and injury is being perpetuated by Louisiana-Pacific and the Forest Service in liquidating old growth forests."[61]

Earth First! and other biocentric ideologies operate on the assumption that agriculture and technological advancements have removed humans from nature and created competing narratives between mankind and nature ever since. Highlighting this assumption, Foreman wrote,

> Before agriculture was midwifed in the Middle East, humans were in the wilderness. We had no concept of "wilderness" because everything was wilderness and we were a part of it. But with irrigation ditches, crop surpluses, and permanent villages, we became apart from the

natural world.... Between the wilderness that created us and the civilization created by us grew an ever-widening rift.[62]

The only way to save the natural world, in Foreman's view, would be for everyone to "follow hunter-gatherers back into a wilderness Eden and abandon virtually everything that civilization has given us."[63] The man-nature dualism is further accentuated by Foreman's insistence that nature's priorities deserve precedence over human priorities, as though it were possible to separate the two. To Foreman, "The preservation of wildness and native diversity is the most important issue; issues directly affecting only humans pale in comparison."[64]

In the late 1970s, the preservationists in the western United States were focusing attention on roadless areas within national forests. The Sierra Club filed a suit in 1972 against the Forest Service for failing to comply with NEPA. The settlement from the lawsuit compelled the Forest Service to conduct wilderness reviews of all roadless areas in national forests.[65] During the 1970s, the Wilderness Society and other environmental groups filed administrative appeals over management decisions by the Forest Service regarding the review of roadless areas and NEPA procedures, and turned to the courts when appeals failed.[66] Facing pressure, the Forest Service launched a second effort to review roadless areas that was labeled RARE II,[67] which "promised a systematic review of the national forest roadless areas that would prevent future lawsuits and end piecemeal wilderness legislation."[68]

Western communities viewed RARE II as a threat to "public lands that they considered to be extensions of their backyards and important to their local economies."[69] During the review for RARE II, citizens of western states overwhelmingly favored land management for multiple-use purposes (62 percent of comments) over absolute protection (7 percent of comments) being sought.[70] Concerned about increasing federal regulations limiting economic growth, groups like the Southern Oregon Resource Alliance and the Las Cruces Jeep Club began arguing against public lands regulations, especially RARE II and the Wilderness Act.[71] The opposition to the Sierra Club, RARE II, Wilderness Act, and federal regulations of western lands became known as the Sagebrush Rebellion. Leaders of the Sagebrush Rebellion argued that, in light of what they considered government overreach, mismanagement, and violations of state sovereignty, the states not the federal government should take control and management of western lands. Although then-presidential candidate Ronald Reagan declared that he was a sagebrush rebel, the movement died a quiet death during his administration.

A renewed opposition to environmental regulations emerged in the 1990s under the banner of wise-use. During and after the Reagan administration, environmentalism became increasingly associated—in conservative circles at least—with "liberals, revolutionaries, and [a] counterculture" that was waging a war against the American West.[72] Rural farmers, ranchers, loggers, and fishers felt they knew how to manage land, having cultivated and worked the land firsthand. Although there is an anti-federal, Sagebrush Rebellion tinge to their rhetoric, they base their claims in constitutional rights to private property and to bear arms as well as populism and citizen involvement. Many in the environmental movement considered and continue to consider western opponents of the mainstream environmental agenda to be "inbred hillbillies" and "chunky cowboys," to use phrases from the comment section of a post on the Southern Utah Wilderness Alliance blog in 2014 and dismiss them as shills for corporate interests. Such reactions, however, show a deep ignorance of the countryside's pent-up resentment toward "meddlesome urbanites and weekend warriors."[73] The environmental agenda has become increasingly perceived in the West as an effort to systematically remove land management decisions from local control. As historian James Turner notes in his seminal article about environmental politics and the response by the political right, "The opposition in the West, represented by the Sagebrush Rebellion and the wise-use movement, emerged as a response . . . to the way environmental decisions were made, the role of the federal government in environmental protection, and the consequences that had for free enterprise and private property owners."[74]

While the Sagebrush Rebellion reflected an unapologetic opposition to the environmental movement in favor of local control and resource industries, the wise-use movement emerged as a more balanced solution to protect ecology and the economy.[75] Wise-use "aimed to reduce the scope of federal government regulations, expand opportunities for economic development, and safeguard individuals' constitutional rights while promoting a 'New Environmentalism' that respected people and nature."[76] Several wise-use-related organizations are influential players in environmental politics and litigation. Among the most visible are the Alliance for America, the Multiple Use and

Inholders Association, Center for the Defense of Free Enterprise, Blue Ribbon Coalition, Mountain States Legal Foundation, Pacific Legal Foundation, and the Western States Public Lands Coalition. These groups have not reversed environmental and public land policy, but they have contributed to an ongoing stalemate over the direction those policies take. The wise-use movement represents more than the political protest of the Sagebrush Rebellion, and it includes more than industry-funded activists. A player in the environmental policy game with an extensive network that includes citizen groups and think tanks, it has learned how to employ many of the same strategies and tactics as the mainstream environmental movement in attempting to impose its view of nature on the American polity.

Conclusions

"Nature undisturbed," also called the balance of nature ideology, has been a useful myth for political entrepreneurs for many decades. It justified and animated a long-term political movement to "preserve" nature, to return much of the United States to an invented state of natural balance, and to create a host of federal laws and subsequent regulations. We have noted that the environmental movement, also founded on balance of nature principles, was given a set of tools to pursue an agenda of elitist preservationism. The federal government, with hastily crafted laws based in myth, basically handed the movement a legislation and litigation toolbox with which to hinder development, commerce, and the use of the country's natural resources.

Recognizing the power of the myth, we assert that it has led to policy dead-ends that do not allow for flexibility and creativity. A belief in a static "balanced" nature ends up with static inflexible laws and policy that do not reflect the reality of nature and human places in it.

As the authors, we pride ourselves on hard-nosed analysis, but we are not immune to the mythology of nature. We spend time each year on wild rivers, in national forests, and in federally designated national recreation areas and wilderness. We partake in the mythology on foot and on horseback, in drift boats and motorboats, and on all-terrain vehicles. As we will see, however, reliance on that mythology has come at great costs to society. Ironically, the costs are borne by the American public, which is the supposed benefactor of both the environmental legislation and the environmental movement's litigation.

5

The Clean Air Act

AIR POLLUTION CAN KILL. In 1948 a thick yellow smog enveloped Donora, Pennsylvania, for five consecutive days. So many people got sick that, as one resident described it, "they had an emergency hospital set up because the local hospitals couldn't handle all the people, and unfortunately in the lower level [of the local hotel] they had a morgue because the funeral directors were overwhelmed with bodies at that time."[1] When the fog finally lifted, eighteen of the town's 14,000 residents had died.

As bad as Donora was, it paled in comparison to what happened to London four years later. "The Great Smog," as it is known today, descended on the city for four days in December 1952. Visibility became so bad that drivers could see no more than a few feet, forcing them to leave their vehicles and walk. The smog was bad enough indoors that, according to Geoffrey Lean, a theater "had to abandon a performance . . . because the audience could not see the stage, and nurses at the Royal London Hospital reported not being able to see from one end of their wards to the other."[2] Ultimately, The Great Smog killed over 4,000 people.

Air pollution has been a problem ever since humans began burning fuel and industrializing. The ancient Romans produced as much lead pollution from lead smelting as the world produced at the beginning of the Industrial Revolution.[3] Air pollution also had a long history in London, beginning in the eleventh century as coal gradually replaced wood as a fuel for both industry and household heating. Over the next few hundred years, various writers would recognize the damaging and unhealthy effects of smoke. Pollution became so severe that by the Middle Ages, the English Parliament enacted bans against coal furnaces.[4] Prohibition, however, combined with high taxes on coal, did little to offset the economic necessity of coal, nor did it alleviate London's stifling pollution. Indur Goklany writes that

> The lack of real alternatives to coal, the existence of more visible and immediate causes of death and disease, which rendered life 'nasty, brutish and short,' and—once the industrial revolution got going—the association of smoke with industrialization and, therefore, jobs and prosperity ensured that there would be little progress in controlling smoke for the next two centuries.[5]

Industrialization increased the burning of coal. In 1800 the United States burned 3 trillion British thermal units (BTU) of fossil fuels. This grew to over 520 trillion BTU in 1860.[6] In an attempt to control their pollution problems, midwestern cities began to pass smoke regulations, but these measures failed to stop both the growth of dirty fossil fuel usage and the subsequent problem of smoke pollution.[7] According to Indur Goklany, "the lack of viable alternatives to coal combustion," combined with several court decisions that overturned various pollution ordinances, prevented the nation's pollution problems from improving, despite outcry from civic groups.[8]

These events were not completely new. Heavy smog had occurred before, but the Donora and London incidents had a great effect on opinions of the American public.[9] After World War II, news media became more national in its audience and more international in its scope.[10] As local incidents were publicized to the wider world, the American public began to demand more pollution control. According to Goklany, "By 1956, there were 82 local air pollution control programs" across the nation.[11]

Often, these ordinances relied on the Ringelmann chart, the first consistent method for measuring pollution levels. The chart was developed around the turn of the twentieth century; users matched the shades of black smoke coming from a chimney against a smoke density chart.[12] New local pollution ordinances contributed to cleaner air, and the court system in the United States was also more favorably disposed to the regulation of business and industry. Regulation made marginal improvements, though as Goklany observes, "it also helped that the regulations did not get ahead of what was economically doable."[13]

The real game-changer was the introduction of cleaner fuels in the form of oil and natural gas. Also, electricity could now be transported over greater distances, which meant that power plants could be located some distance away from population centers. The trend of replacing wood and coal with oil and gas continued until "the smoke problem was virtually solved in most urban areas" by the 1960s. Nevertheless, regulators followed up, imposing stricter standards to match the improving air quality.[14]

Not only did the air improve without sacrificing economic growth, the air improved *because* of economic growth—business and industry had strong incentives to improve efficiency, which led to advances in technology, which led to cleaner energy. Although it's easy to decry the horrendous pollution of the early Industrial Age, those early, dirtier technologies made it possible for the economy to progress toward cleaner energy.

Cleaner, however, is a relative term, and while oil and gas did not emit the heavy black smoke that coal did, they still released sulfur dioxide and particulate matter, both of which are dangerous in high concentrations. When sulfur dioxide and particulate matter became a serious problem in the Los Angeles area of California, regulators worked to control emissions from factories, power plants, and other stationary sources. While they did so with some success, their efforts did little to solve the smog problem. Automobiles were becoming more common and numerous in the 1950s, and carbon monoxide, a byproduct of combustion, was threatening the health of Los Angeles residents, along with a new type of pollution: photochemical smog. This secondary pollutant is formed when hydrocarbons and nitrogen oxides released from automobiles react with sunlight to form tropospheric ozone and other oxidants.[15] In another case of innovation preceding regulation, by 1961 auto manufacturers were building new automobiles with devices that prevented the venting of hydrocarbons from the crankcase into the atmosphere.[16] This improved the life of engines while also cutting down on smog.[17] As a part of its new motor vehicle emissions standards imposed in 1966, the state of California required crankcase devices to be installed in new cars sold in the state.[18] In other words, business and industry again addressed their own problems by implementing better technology.

Until 1955 air pollution control was a local matter. The Air Pollution Control Act of 1955 was the federal government's first attempt to deal with air pollution.[19] Although the legislation was federal, the government simply provided funding for research into the sources of pollution and new techniques to monitor and control it.[20] It was not involved in implementation of any requirements. State and local air quality efforts were already having substantial effects by the time the federal government got involved.[21]

Eight years later, Congress passed the Clean Air Act of 1963. It was the first time the federal government had attempted to implement air pollution control directly.[22] Even so, the control was not a broad-based emissions requirement. Instead the law established a case-by-case procedure for dealing with pollution problems. When requested by state or local agencies, the U.S. Department of Health, Education, and Welfare could organize a conference with the interested parties, evaluate the pollution situation, and determine if further regulatory action was needed.[23] The power to regulate was fairly limited. If the federal government determined that regulation was needed, then it would use the court system to enforce the decision, and the federal government would regulate intrastate pollution if the state's governor requested it (Anderson, 2000, p. 278).[24]

The Air Quality Act of 1967 authorized the federal government to establish acceptable ambient air quality levels. The secretary of health, education and welfare was to designate air quality control regions, while states were to develop state implementation plans and submit them for approval. Only about one-third of the expected air quality control regions were designated, and no state plan was ever approved. Despite the failure of the 1967 legislation, air quality continued to improve, probably because of state and local regulations and cleaner production technologies. By the 1970s, the nation had reached a point where high-fatality smog events like those in Donora and London were a thing of the past.

Pollution levels were declining, but that was not the public perception about air quality. Senator Edmund Muskie, who was the Democratic nominee for vice president in 1968 (and who expected to run for the presidency in 1972), and President Richard Nixon were competing to see who could be viewed as the best environmentalist. According to a Nixon aide,

> Yet there is still only one word, hysteria, to describe the Washington mood on the environment in the fall of 1969. The words pollution and

environment were on every politician's lips. The press gave the issue extraordinary coverage, Congress responded by producing environment-related bills by the bushel, and the President was in danger of being left behind.[25]

Nixon responded by proposing a bill that, after a few revisions by Congress, would become the basis for the Clean Air Act of 1970 (CAA).

The CAA substantially amended the original Clean Air Act of 1963, and laid a foundation for pollution control that future amendments would modify. The most important change was the establishment of National Ambient Air Quality Standards (NAAQS), which would be administered by the Environmental Protection Agency (EPA).[26] NAAQS, levels of pollution deemed acceptable by the EPA, were set for six pollutants: carbon monoxide, tropospheric ozone, sulfur dioxide, particulate matter, nitrogen dioxide, and hydrocarbons. Lead was added in 1976, and hydrocarbons were removed in 1978 because they were necessary for the creation of tropospheric ozone, and therefore covered by that standard.[27]

Under the CAA, areas of the country that fail to achieve the standard set by the EPA are designated as nonattainment areas. States are required to submit a state implementation plan (SIP) to the EPA for approval, detailing how the area will improve its air quality and meet the NAAQS. If a state fails to complete such a plan, the EPA will impose its own plan and can deny the state federal funds. The potential denial of federal funds acts as a strong incentive to comply with the EPA.

Another new development included in the CAA was the mandate of new source performance standards, which set emissions caps for stationary sources, including factories, power plants, and even wood stoves. New sources of pollution were required to comply with the emissions cap.[28] Mobile sources were also affected by automobile emissions regulations. While the auto industry had been in favor of a national emission standard because it was easier to comply with a single standard than standards that changed from state to state, they were now faced with regulations that "could not be met with existing technology."[29]

Rather than building on the existing bottom-up and state-led system, the CAA imposed a top-down, bureaucratic approach. Under the CAA, states had to receive approval from the federal government for their control policies. State public officials were now managed by federal bureaucrats.[30]

As a result, the CAA satisfied neither environmentalists nor polluters. Industry pushed back against a law they saw as going too far, and many polluting industries, like automakers and energy producers, successfully influenced bureaucrats into writing softer implementation rules. Environmentalists thought the law failed to adequately protect air quality, and they sought to both expand the CAA's scope and strengthen its regulations against industry. The courts were one of the tools environmentalists used to accomplish their goals. In a 1972 lawsuit brought by the Sierra Club, a district court ordered "the EPA to create a program for preventing the deterioration of air quality in regions where it was above the NAAQS levels."[31] In other words, those areas that were already doing better than the NAAQS would not be allowed to deteriorate, even if after deterioration they still met the NAAQS. This was called Prevention of Significant Deterioration (PSD).[32]

The Clean Air Act amendments of 1977 were passed in response to these developments. They gave the court-ordered PSD program a legislative mandate and, in the process, revised and expanded it with tighter restrictions. The 1977 amendments also revised many of the regulations that the CAA had imposed on automakers and other polluters. The increased complexity of the 1977 amendments created an opportunity (especially among eastern coal producers) for "rent seeking," which is a term we will cover in greater detail later in this chapter.

Ultimately, according to Andrew P. Morriss, the 1977 amendments "produced a mixed result for all sides—automakers succeeded in delaying and weakening standards, eastern coal interests won big, and environmentalists did well in preserving and extending the PSD program."[33] Bruce Ackerman and William T. Hassler go one step further. In their book *Clean Coal Dirty Air*, they analyze the politics of the 1977 amendments and conclude that they were really a "multi-billion dollar bail-out for high sulfur coal producers."[34]

Over the next thirteen years, these interests would continue to influence lawmakers to modify the CAA. According to Morriss, "The maneuvering through the 1980s pitted regional, industrial, environmental, bureaucratic, and other interest groups against each other in shifting coalitions that produced repeated failed efforts to amend the CAA."[35] Each group was attempting to modify the CAA for different reasons. The EPA recognized that SIPs lacked good data and often relied on slapdash analysis in order to meet deadlines.

After a convoluted process of tactical political machinations, Congress passed the Clean Air Act Amendments of 1990. The 1990 amendments introduced several new provisions that greatly increased the complexity of pollution control regulations. The nonattainment provisions received "an elaborate new classification system." This meant stricter emissions requirements and more inspections for automobiles. Various toxic pollutants, which had been ignored in the previous amendments, were the subjects of a new, complicated regulatory structure. Acid rain was tackled by an international protocol. A cap-and-trade system was established, allowing polluters to sell or trade sulfur dioxide emissions permits.

From the original version of the CAA to the 1990 amendments, growing federal programs incrementally crowded out state and local government action. Prior to the centralization of air pollution control, state and local governments had been quite successful at regulating pollution. There were some impressive early successes under the emissions trading system allowed under the 1990 amendments. However, those gains have been eroded through time. Overall, the CAA and its amendments have been less successful at actually cleaning the air than at allowing the EPA to extend its reach.

Massachusetts v. Environmental Protection Agency

Perhaps no case illustrates how the CAA has led to bureaucratic overreach as well as *Massachusetts v. Environmental Protection Agency*. This 2007 U.S. Supreme Court case tasked the EPA with regulating greenhouse gases emitted by automobiles, extending the agency's jurisdiction for the first time to the mitigation of global warming.[36] The court's decision in this case dramatically expanded the authority and jurisdiction of the EPA. The EPA was no longer tasked solely with regulating pollutants that might "reasonably be anticipated to endanger public health or welfare," as dictated by the CAA.[37] Instead, the EPA was required to make a decision as to "whether greenhouse gas emissions contribute to climate change" and to regulate them accordingly.[38]

In 1999 the State of Massachusetts, along with eleven other states, two cities, the District of Colombia, American Samoa, and thirteen private environmental organizations, petitioned the EPA to regulate the emissions of greenhouse gases from new motor vehicles.[39] The EPA denied this petition in 2003, arguing that the CAA did not give the agency the authority to regulate greenhouse gas emissions. When it was originally passed, the CAA was not intended to mitigate greenhouse gases like carbon dioxide. As described above, the CAA was meant to regulate pollutants that could be reasonably expected to damage public health or welfare. Carbon dioxide and other greenhouse gases do not cause a direct harm to public health. John Copeland Nagle described how carbon dioxide itself does not harm human health, saying, "Unlike most air pollutants, CO_2 occurs naturally in the atmosphere, is actually necessary for human life."[40]

In justifying its denial of the petition, the EPA also argued that it would delay its decision on whether to regulate greenhouse gas emissions until more conclusive research addressed "the causes, extent and significance of climate change and the potential options for addressing it."[41] In other words, the EPA was not prepared to classify carbon dioxide and other greenhouse gases as harmful to the public. Because these gases occur naturally and do not threaten human health, such a classification would have to assume that climate change is occurring, that it is being caused by greenhouse gas emissions, and that these emissions are causing a deterioration of public health. Those are major assumptions. Massachusetts appealed to the D.C. Circuit Court Appeals, which ruled in favor of the EPA.[42]

Establishing Standing

In 2006 the Supreme Court granted the case certiorari, agreeing to reopen arguments once again. The stage was set for the case to become a central part of the controversial climate change debate. Two main questions lay before the court. The first was "whether EPA has the statutory authority to regulate greenhouse gas emissions from new motor vehicles," and the second was "whether its stated reasons for refusing to do so are consistent with the statute."[43] Before it could answer either of these questions, however, the court had to decide whether Massachusetts had the right to bring the case to court in the first place. In legal terms, the court had to determine whether Massachusetts had standing.

In determining standing, the court relied on *Lujan v. National Wildlife Federation,* which established a three-part test for determining standing. According to the Lujan case, to establish standing, "a litigant must demonstrate that it has suffered a concrete and particularized injury that is either actual or imminent, that the injury is fairly traceable to the defendant, and that it is likely that a favorable decision will redress the injury."[44]

To meet the first part of the Lujan test, the court found that Massachusetts' "concrete and particularized injury" consisted of rising sea levels caused by global warming. If global warming continued unchecked, then Massachusetts' coastline could be "either permanently lost through inundation or temporarily lost through periodic storm surge and flooding events."[45] The costs of such property loss and damage "could run well into the hundreds of millions of dollars."[46] Thus Massachusetts overcame the first hurdle to achieving standing.

Next, the court had to decide whether the Environmental Protection Agency's (EPA's) refusal to regulate greenhouse gases caused Massachusetts' injury. In other words, is the threat of rising sea levels the EPA's fault? Making this connection seemed like quite a stretch, but it was one the court was willing to make. The Supreme Court found that causation did exist because "judged by any standard, U.S. motor-vehicle emissions make a meaningful contribution to greenhouse gas concentrations and hence, according to petitioners, to global warming."[47] If the EPA refused to regulate emissions from automobiles, it was contributing to global warming. Because this would lead to higher sea levels, the EPA was heightening the threat to Massachusetts' coastline.

Finally, the court had to decide whether a solution existed that would "redress the injury." In other words, if the court were to decide against the EPA, could it impose a punishment that would make up for the damage and loss of property that Massachusetts had allegedly suffered? The court stated in its opinion that, "While it may be true that regulating motor-vehicle emissions will not by itself *reverse* global warming, it by no means follows that we lack jurisdiction to decide whether EPA has a duty to take steps to *slow* or *reduce* it."[48] The court decided that if Massachusetts were to achieve the "relief they seek," then "the risk of catastrophic harm" would be reduced.[49] Massachusetts had overcome the Lujan test, with a little help from a Supreme Court willing to interpret concepts like *harm* and *cause* as widely as necessary.

The Court's Decision

Once the Supreme Court had determined that Massachusetts had standing to sue, the next question on the table was "whether §202(a)(1) of the Clean Air Act authorizes EPA to regulate greenhouse gas emissions from new motor vehicles in the event that it forms a 'judgment' that such emissions contribute to climate change."[50] It seemed to some that the Court was determined to find that Massachusetts had standing in the case and equally determined to interpret the law as widely as possible to find the EPA at fault.

The Supreme Court began its decision by citing section 202(a)(1) of the CAA, which states that the EPA must regulate "any air pollutant from any class or classes of new motor vehicles or new motor vehicle engines, which in [the Administrator's] judgment cause, or contribute to, air pollution which may reasonably be anticipated to endanger public health or welfare."[51] As discussed earlier, the CAA was not intended to regulate substances like carbon dioxide because they do not directly lead to the endangerment of public health or welfare.

The Supreme Court preferred to take a wider interpretation of the CAA's provisions, calling the law's definition of air pollutants "sweeping" and "capacious."[52] Although the court admitted that, in passing the CAA, the legislators "might not have appreciated the possibility that burning fossil fuels could lead to global warming," they also asserted that the law was meant to be flexible to avoid obsolescence.[53] The court held that greenhouse gases fit "well within" the CAA's definition of an "air pollutant" and that the EPA was thus required to regulate such emissions.[54] With that determination, the EPA had lost the first part of the battle.

The second part of the case hinged on the question of whether the EPA's reasons for refusing Massachusetts' petition were consistent with the law. The Supreme Court argued that, in order to be consistent with the CAA, the EPA could refuse to regulate greenhouse gas emissions only "if it determines that greenhouse gases do not contribute to climate change or if it provides some reasonable explanation as to why it cannot or will not exercise its discretion to determine whether they do."[55]

The EPA's reasons for not regulating greenhouse gas emissions from new vehicles were many. For example, the EPA argued that many voluntary programs already existed that effectively responded to the "threat of global warming."[56] The agency also argued that if it were to regulate greenhouse gases, this might make it more difficult for the president of the United States to negotiate with developing nations on reducing emissions. Finally, the EPA stated that regulating vehicle emissions would entail "an inefficient, piecemeal approach to address the climate change issue."[57] The EPA did not want its regulation to interfere with the comprehensive climate change strategy of the executive branch. The Supreme Court, however, did not find these to be "reasonable" explanations for refusing to regulate greenhouse gas emissions. The court thus found the EPA's actions to be "arbitrary, capricious . . . or otherwise not in accordance with law,"[58] and Massachusetts had won its case against the EPA.

The Aftermath of Massachusetts v. Environmental Protection Agency

After the court issued its finding in *Massachusetts v. Environmental Protection Agency,* the EPA was left with three options. First, it could issue a finding that greenhouse gases could be reasonably anticipated to endanger public health or welfare. Second, the agency could issue a finding of no endangerment from greenhouse gases. Finally, the agency could provide an explanation for why it refused to do either.[59] The EPA chose the first option. It issued a finding of endangerment in 2009. This decision, however, led to a dramatic expansion of the agency's reach and scope. To deal with this expansion, the EPA attempted to narrow the endangerment finding with what is now known as the Tailoring Rule.

The Rise and Decline of the Tailoring Rule

The most direct result of *Massachusetts v EPA* is that the EPA was required to regulate greenhouse gas emissions from new automobiles. In 2010 the EPA and the U.S. Department of Transportation issued the first national rule limiting greenhouse gas emissions and setting fuel economy standards for cars and light trucks for model years 2012 through 2016.[60] Because the case dealt directly with emissions from cars and their contribution to global warming, this result made sense.

While the case centered on emissions from vehicles, it also required the EPA to reach a decision about whether greenhouse gases endanger public health in general. In its 2009 endangerment finding, the EPA declared "current and projected concentrations of the six key well-mixed greenhouse gases . . . in the atmosphere threaten the public health and welfare of current and future generations."[61] Upon establishing the link between greenhouse gas emissions and public health, the EPA unilaterally expanded the Supreme Court decision beyond automobiles to all sources of greenhouse gases. Recognizing the impossibility of the task, the EPA created the Tailoring Rule to limit the scope.

The Tailoring Rule limited permitting requirements to the largest emitters of greenhouse gases, including power plants, refineries, and cement production facilities, while excluding small businesses like farms, restaurants, and commercial facilities.[62] Starting in January 2011, existing facilities, which already had to undergo permitting for other pollutants, would be phased in to greenhouse gas permitting requirements, only if they increased emissions by at least 75,000 tons per year (tpy).[63] Beginning in July 2011, new sources that emitted at least 100,000 tpy of carbon dioxide would also be phased in.[64] Facilities emitting less than 50,000 tpy of greenhouse gases are not anticipated to be subject to regulation prior to 2016.[65] As Nathan D. Riccardi notes, "This is a significant relaxation of the normal statutory threshold of 250 tpy of any air pollutant."[66]

The legal justifications for the Tailoring Rule are the so-called "Chevron Defense," which arose from *Chevron U.S.A., Inc. v. Natural Resources Defense Council, Inc.*[67] and the "absurd results doctrine." The Chevron Defense established a two-prong test for whether an agency's interpretation of a law is valid. The first prong says that if the intent of Congress is clear, the agency must act accordingly. The second prong allows for an agency to "employ a permissible interpretation of the statute" if the intent of Congress is unclear or if it has not addressed the issue at hand.[68]

Although the language of the CAA clearly states that emissions classified as harmful pollutants should be regulated at the 250-tpy level, the EPA claimed that this literal interpretation would lead to absurd results if applied to greenhouse gas emissions.[69] The EPA said that a literal application would require millions of new applications for permits, overwhelming the agency to the point where it could not effectively operate. The EPA would not be able to do its job, so it invoked the absurd results doctrine, claiming that "where a literal reading of a statutory term would lead to absurd results, the term simply has no meaning."[70] The agency then claimed the second prong of the Chevron Defense should apply: The EPA should be allowed to use its judgment in determining how to interpret and apply the statute.

A challenge to the Tailoring rule made it to the Supreme Court as *Utility Air Regulatory Group v. Environmental Protection Agency*. The decision in June 2014 struck down the Tailoring Rule, while reaffirming EPA's authority to regulate greenhouse gases under the CAA. The majority opinion said the EPA is not allowed to revise clear statutory text in order to achieve agency policy goals. Doing so is a violation of the separation of powers. Thus, "tailoring" is gone but *Massachusetts v. Environmental Protection Agency* remains intact.

Massachusetts v. Environmental Protection Agency changed the interpretation of the CAA dramatically. The Supreme Court essentially decided that carbon dioxide and other greenhouse gases that are necessary for human life should be regulated just like toxic gases. This enabled the EPA to unilaterally decide whether the broad contribution of greenhouse gases to climate change was impacting public health and to regulate them accordingly.

Two fundamental problems at the heart of the CAA were enhanced by *Massachusetts v. EPA*. The first is the inability of the EPA to possess the knowledge or ability to tackle all of the problems now deemed to require regulation by the expanded CAA. The second is that the expansion also increases the complexity of the CAA, which augments the ability of interest groups to engage in rent-seeking behaviors. In the case of the Tailoring Rule, the EPA attempted to expand the scope of the CAA beyond what a majority of the Supreme Court believe are reasonable interpretations of its original intent. It was an attempt to warp the CAA sufficiently for it to become a catchall for regulations to reduce global climate change.

The Knowledge Economy

Economist Friedrich Hayek astutely noted that the problem of how to use resources in society is actually "the problem of the utilization of knowledge," which by its nature "is not given to anyone in its totality."[71] In his essay, "The Use of Knowledge in Society," Hayek further analyzed the importance of knowledge in economic systems. Because knowledge is essential to planning and making economic decisions, and because knowledge is inherently "dispersed among all the people," Hayek concluded that the best form of planning for an economy is by the means of competition, which is "decentralized planning by many separate persons."[72] Applying Hayek's analysis to the CAA illuminates many of the dangers in the extension of its power accomplished by *Massachusetts v. Environmental Protection Agency*.

The justifications for bureaucratic overreach arise from faulty premises. Congress empowers bureaucracies based on the assumption that complicated decisions are best left to those with the most knowledge and that they will always choose experts to fill bureaucratic positions. Their decision-making power is based on their assumed knowledge. This assumption is faulty, however, for two reasons. First, it fails to demonstrate that Congress will choose those with the right knowledge base. Second, knowledge, as Hayek observed, is by its nature decentralized. If decision-making power is to be based on knowledge, then it should follow that it, too, ought to be decentralized. The expansion of the CAA and the EPA's regulatory responsibilities under the 2007 Supreme Court decision only enhances the problems of knowledge and power inherent in bureaucracies.

This false assumption is at the root of modern approaches to bureaucracy and government. Angelo Codevilla, a professor at Boston University, notes that the modern idea of bureaucracies stems from the Age of Enlightenment in the eighteenth century. As thinkers began to question the inherited political and social organizations of continental Europe, they began to search for new systems. Many held that there were rational solutions to the organization of society and that if those with the most knowledge were put in positions of power, they could organize society according to the principles of reason. This idea of "enlightened despotism" led to the formation of the first modern bureaucratic system in Napoleonic France.[73]

The Enlightenment model of bureaucracies is essentially technocratic central planning. It remains with us today, expressed perfectly in the expansion of the power of the EPA under the CAA. Angelo Codevilla criticizes this model, noting the inherent problems it creates. "How does one ascend the heights of technocracy?" Codevilla asks. "Which degree must they hold? How much knowledge must they demonstrate? If the person appointing them to the position does not themselves hold the knowledge, then how can we judge if they are the adequate judge to judge who should be in the position?" Codevilla further observes that "scientists are not immune to groupthink, to self-interest, to dishonesty, to mutual defense or antagonism, never mind to error."[74]

Codevilla's observation is strikingly similar to the findings of behavioral economists who examine the psychology of decision making. Nobel Prize-winning behavioral economist Daniel Kahneman published a summary of his research in the field in 2011 entitled, "Thinking, Fast and Slow," in which he demonstrated the importance of research in the field of behavioral economics for organizations.[75] His conclusions are directly relevant to our analysis of the overextension of the EPA's regulatory power because they demonstrate the weaknesses of centralized decision-making on complicated issues.

Kahneman's research demonstrates that the human mind simplifies decision-making through the application of biases and heuristics, which are procedures that simplify the thinking process. While Kahneman and his associates have discovered many heuristics, among the most powerful and the most relevant to the current discussion is the cognitive bias toward the present, a simple heuristic in which the mind discounts the unavailability of information. Decision makers within organizations often ignore the fact that they do not see the whole picture. References to this heuristic in his work are so ubiquitous that he coins an acronym to describe it—WYSIATI, "what you see is all there is." The problem, Kahneman demonstrates, is that people "cannot help dealing with the limited information [they] have as if it were all there is to know." He therefore concludes that "our comforting conviction that the world makes sense rests on a secure foundation: our almost unlimited ability to ignore our own ignorance."[76]

The writings of both Kahneman and Codevilla force us to deeply question the centralization of decision-making power in the hands of bureaucratic organizations, no matter the claims to expertise or knowledge. No single person can know all there is to know on a subject, and the centralization of the decision-making power in a context where knowledge is decentralized will serve only to enhance groupthink and other similar biases and heuristics. When decision-making power is concentrated in the unchecked hands of a few so-called experts, they will be more prone to WYSIATI thinking, assuming that they see the whole picture when it is demonstrable that they do not. In addition, the present heuristic will cause bureaucrats to consider the political and situational incentives of the present above searching for unknown scientific variables to get a more complete vision.

In fact, the problem has been demonstrated with regard to the scientific approaches of the EPA. In 1995 Ken Sexton of the University of Minnesota published an article in the journal *Environmental Health Perspectives,* showing that changes in the methods used by the EPA to measure hazardous air pollutants did not reflect scientific consensus on the issue.[77] Because decision-making power is centralized, however, it doesn't matter, and the decisions are made anyway. Sexton further notes, "the more significant situational variables are the age of the relevant public policy issue . . . . and bureaucratic realities of real-world pressures and demands on regulatory decision makers."[78]

The economic consulting firm NERA has also examined the science behind the EPA's approach to measuring health benefits from reductions of particulate matter$_{2.5}$, showing that they have blatantly ignored important variables and discounted valuable information in their evaluations.[79] Similar research has shown that the EPA consistently overlooks important variables and twists the science to match its political agenda.[80] Geoffrey Kabat, a scientist working in California, wrote in *Forbes Magazine* that during the public comment period for one new EPA regulation, "a number of scientists (myself included) wrote in to correct errors and omissions in the draft report. However, there [was] no evidence that critical comments were given any attention by the agency."[81] With the bureaucracy blinded by biases, WYSIATI ruled the day.

As the powers of the EPA have grown larger after the decision in *Massachusetts v. Environmental Protection Agency*, the problem of knowledge has grown exponentially. To tackle a problem as amorphous as global climate change, the EPA would need to have an intimate knowledge of every source of emissions in the United States and every economic factor involved in its production. The costs of being wrong on such a large scale could be economically devastating. The EPA simply does not have the regulatory capacity to singlehandedly tackle such a large issue, and it is evident that the EPA is aware of this fact—it is what led to the development of the Tailoring Rule. The Tailoring Rule, however, cannot change or compensate for the fact that the EPA's regulatory vision far surpasses its capacity to know and understand all of the issues at hand.

The solution to the problem of incomplete knowledge does not come from endowing with decision-making power those whom Congress deems to be experts, but rather by enhancing the competitive nature of the knowledge economy. At the end of his book, Kahneman concludes that "a world in which firms compete by offering better products is preferable to one in which the winner is the firm that is best at obfuscation."[82] He likewise observes that "an organization is a factory that manufactures judgments and decisions."[83] If we follow Kahneman's line of reasoning and view the EPA as a factory that manufactures judgments and decisions about environmental problems, it is clear that those judgments are produced not by a competitive process, but rather in obfuscation.

Environmental situations and circumstances differ not only from state to state, but from locale to locale. The dynamics of energy markets differ across the nation. Scientific knowledge is not centralized but found in the minds of many different scientists who participate at different levels of society. Each of them is vital to the whole, and Hayek's observation that an economy planned by many separate persons at different levels is most efficient is highly relevant to conversations about air pollution and the CAA. Increased centralized power of the EPA only damages the process of searching for the right approach to dealing with complicated economic issues.

Even more potent than the problems inherent to centralization in the economy of knowledge, however, are those observed by Sexton: that bureaucrats play an essentially political role and respond to incentives. Because both legislatures and the bureaucracies they create are inescapably caught up in the ecology of politics, the policies they produce cannot be untangled from the lobbying efforts of interest groups. The problems of bureaucracy and the economy of knowledge are created by the increased complexity of the law. The complexity of the law is part of a feedback loop with interest groups caused by rent-seeking behaviors.

Rent-Seeking

As was demonstrated in our discussion of the CAA's history, viewing the CAA as the causal factor in the vast decrease in air pollution in the last forty years is rather dubious. That the CAA is vastly inefficient and overly complicated is no secret; as early as 1981, Deputy General Counsel of the EPA, William F. Pederson, declared in a *Pennsylvania Law Review* article entitled "Why the Clean Air Act Works Badly" that the act was "too cumbersome, and ... poorly suited to address ... emerging issues in air pollution."[84] Instead, technological advances, an increase in wealth, social consciousness, and state and local regulations are more likely to be the causal factors. The question we must ask then is what is the impetus behind the CAA? If it has not been effective, then why has it been used to expand the powers of the EPA? What factors led to the incredibly complex network of laws and legislation that regulate pollutants and emissions in the United States?

Many Americans view environmental policy as being immune from the special-interest lobbying that plagues other political problems. At the time of the passage of the CAA, the environment was "the favorite sacred issue of all politicians, all TV networks, all writers, all good-willed people of any party."[85] Political ecology, however, is in no way immune from the ecology of politics. The generally accepted story behind the CAA tends to be one of goodwill—that benevolent lawmakers wanted to advance the public good by designing and passing the CAA. However, a review of the role of interest groups in the formation and evolution of the CAA reveals that rent-seeking behaviors are most likely the primary impetus behind the CAA and its continued development. The CAA was born not of goodwill, but of the ecology of politics.

The lobbying activities of interest groups and the complexity of the law form a feedback loop. Interest groups lobby for certain aspects of the law to be changed, altered, or reinterpreted, which creates greater complexity, which in turn increases the number of interest groups with incentives to lobby for changes in the law. As Morriss has noted, "the complexity seems to exist to hide special favors for regions and industries with powerful legislative advocates, such as West Virginia's high sulfur coal region and Detroit's auto manufacturers."[86] Far too often, narrow political and economic interests become the focal point of developments in environmental law and the impetus that moves its evolution forward.

This feedback loop is particularly evident in the events that led to the Supreme Court's hearing of *Massachusetts v. Environmental Protection Agency*. On October 20, 1999, a group of nineteen private organizations filed a petition to the EPA. Their argument was that section 202(a)(1) of the CAA mandated the regulation of greenhouse gas emissions from new motor vehicles. Reasoning along the same lines as those in the Massachusetts case, the petitioners held that greenhouse gases qualified as a pollutant that threatened public health and welfare because of their contribution to global climate change.[87]

The petitioners knew that if greenhouse gases were classified as pollutants that harmed public health and welfare, the EPA's regulatory power would quickly extend beyond automobile emissions. Unsurprisingly, many of the petitioners represented industry interests that would gain greatly if the EPA had the power to regulate carbon emissions. Applied Power Technologies, Inc., Biofuels America, the New Mexico Solar Energy Association, the Solar Energy Industries Association, and the SUN DAY Campaign were among the nineteen petitioners.[88] These groups represent industries as varied as natural gas, solar energy, and ethanol production, each of which would stand to profit economically (and thus "pay the rent") if the EPA's power to regulate carbon emissions was increased and extended. The coal industry, on the other hand, would suffer immensely. This 1999 petition started the process that led to the 2007 Supreme Court decision requiring the EPA to assess whether or not greenhouse gases threatened public health and welfare, a decision that was not simply an environmental mandate but an active picking and choosing of winners and losers in the energy market.

The profound entanglement between interest groups and the CAA extends back to its very formation. In the first seven years of the CAA's existence, it became the battleground for the coal industry.[89] Coal mined in the eastern United States has much higher sulfur content than that of the west and produces dirtier emissions when it is burned. The original CAA regulated coal by its sulfur content, setting the standard of 1.2 pounds of sulfur dioxide per million BTUs of coal. While the standard could be met by the installation of air scrubbers, most plants found these too expensive, and it was cheaper to transport cleaner-burning coal from the western United States than to purchase the equipment.[90]

In 1977 Congress amended the CAA and changed the industry standard. After vigorous lobbying by an ironic combination of environmental lobbyists and groups representing eastern coal producers, the standard was changed from regulating the amount of sulfur dioxide allowed in emissions to a blanket mandate that all coal-fired power plants install scrubbers.[91] This changed the dynamics of the coal market, erasing the comparative advantage of western coal that was created by the standards of the 1970 CAA. Furthermore, eastern coal producers joined with the Sierra Club to lobby Congress to require the use of local coal in eastern power plants, which legislatures then encoded in Section 125 of the CAA. In effect, the 1977 amendments to the CAA became a bailout for high-sulfur coal producers.[92] Senator Edmund Muskie summarized the nature of the amendments on the floor of the U.S. Senate when he observed that "the dominant thrust of this amendment is not its relationship to clean air, but its relationship to the economics of the areas it is trying to protect."[93]

Jonathan Adler, an environmental law expert at Case Western Reserve University, notes that one of the ironic effects of amendments with regard to coal standards was to "extend the life of older, otherwise obsolete, coal-fired plants." By installing scrubbers, many old plants were able to avoid being shut down when they would have otherwise been replaced with newer plants that used more environmentally friendly technologies. Older plants were now more cost-effective in comparison to the construction of new plants, prolonging the life of environmentally unfriendly technologies and practices. According to Adler, "some regions of the country actually saw an increase in sulfur dioxide emissions as a result of the new law, and the amount of scrubber sludge requiring disposal increased substantially."[94]

As we review the history of the CAA, the idea that goodwill played a role in its conception is increasingly absurd. The EPA has acknowledged the role of rent-seeking and interest groups in the formation of policy. Former Deputy Administrator A. James Barnes noted that attempts "to gain a competitive advantage through manipulation of the regulatory process" were "occurring with increasing frequency."[95] Industry interest groups frequently lobby together with environmental groups to achieve regulations that benefit their corner of the market. As bureaucratic power extends, this phenomenon only increases.

Conclusion

Although data do not show conclusively whether the CAA and its expansion have had a profound effect on air pollution, the regulations have caused a negative impact on the economy. Rent-seeking behaviors have allowed certain industries to gain a competitive advantage over firms that may have otherwise been more cost-effective in a fair market. The ultimate cost of this inefficiency is born doubly by the consumer, who bears both the tax burden of the increase in regulators and also the cost burden of more expensive forms of energy.

Michael Greenstone of the University of Chicago has demonstrated that in its first fifteen years, the CAA caused the loss of 590,000 jobs in counties and states labeled nonattainment by EPA standards. Regulations further led to the loss of $39 billion in capital stock and $75 billion in output in industries judged to be pollution intensive.[96] This research was not comprehensive, and it is expected that these trends would extend to states and counties that reached attainment. The actual quantity of economic loss in the first fifteen years of the CAA was likely much higher than the numbers reported by Greenstone.

Current expansions of the CAA pose equal risks to local economies. An especially pertinent example is the case of Pennsylvania. Governor Tom Corbett of Pennsylvania has labeled the EPA's most recent expansion of regulation on carbon emissions, the result of *Massachusetts v. Environmental Protection Agency,* a "war on coal and a war on jobs." He said, "Here in Pennsylvania, nearly 63,000 men and women work in jobs supported by the coal industry." Corbett called the Obama proposal "not only a war on coal, as suggested by a White House climate advisor, but also a war on jobs."[97] The most recent standards nearly eliminate any competitive advantage of the coal industry in the market, while it boosts the advantage of those involved in the natural gas industry. Essentially, the EPA has chosen the winners and losers in the energy markets, with tens of thousands of American jobs at stake.

6

The National Environmental Policy Act

IN 1989, in an effort to meet a critical landfill capacity shortfall in southern California, Kaiser Ventures, Inc., filed an application to exchange land with the U.S. Bureau of Land Management (BLM) and construct a new landfill. Because the project involved federal land, the project proponents were required to produce an environmental impact statement under the National Environmental Policy Act (NEPA). The project was expected to meet the shortage of landfill areas, operate for 117 years, support or save 1,354 jobs annually, and meet or exceed the guidelines set by the U.S. Environmental Protection Agency (EPA).[1]

After twenty years of NEPA process and 50,000 pages of bureaucratic paperwork, project opponents successfully litigated and stopped the project.[2] The successful litigation hinged on NEPA's "reasonable alternatives" requirement, which is discussed in detail later in this chapter. The outcome demonstrates NEPA's failure to achieve its statutory goals or to "encourage productive and enjoyable harmony between man and his environment," as the law intended.[3] This story is just one example among many of how NEPA causes delays and increases project costs for the benefit of a small group of political entrepreneurs. Ultimately, taxpayers end up paying many of these costs and suffer from slowed economic development in the process.

NEPA is considered the Magna Carta of environmental law and the foundation for modern environmental policy.[4] It is also described as "[a] boondoggle," "inefficient," "expensive," "[a] byzantine bureaucratic maze that stifles productivity," and "[a] blunt instrument of project elimination."[5] This chapter will explore how NEPA began as the Magna Carta of environmental law but became the "blunt instrument" that it is today—a policy used by environmentalists and project opponents to obstruct development, no matter the cost.

Members of the public often misinterpret the acronym NEPA to mean National Environmental *Protection* Act, when the acronym actually stands for the National Environmental *Policy* Act. This misinterpretation applies not only to the title of the act but also to its intended purpose. The law actually offers little substantive protection for the environment. Any protections that occur are often side effects of legal-action threats and social coercion used by those who oppose a project.[6]

Although NEPA was originally predicated on the flawed "balance of nature" theory, its primary contemporary use (and misuse) lies squarely in the realm of political entrepreneurship. Since its passage, NEPA has been incrementally bastardized by agencies and organizations to build entrepreneurial capital. That is, by complying with and manipulating the NEPA process, bureaucrats and activists can convince higher-ups and constituents that their positions are increasingly relevant and require more funding.

Today, NEPA has become a favorite tool of political entrepreneurs, who use the law to subvert, obstruct, delay, and drive up costs for projects. In what follows, we will first analyze the background and intended purpose of NEPA. We will then discuss how NEPA was flawed from the beginning because of its basis in the debunked balance of nature theory. Finally, we will discuss the specific methodologies political entrepreneurs use to manipulate the law to achieve their ends, illustrating the need for reform of NEPA.

How NEPA Works

Throughout the 1950s and 1960s, public concern for the protection of the environment grew steadily in the United States. Increasingly, people became aware of and concerned about the impact that humans were having on their environment. During the 1970s, Congress reacted to this concern by passing major environmental laws, including the Clean Air Act and Clean Water Act. On January 1, 1970, President Richard M. Nixon signed NEPA into law, declaring environmental protection a national policy.[7] NEPA, however, was substantially different from other environmental laws passed during this same time period. The Clean Air Act and Endangered Species Act, for example, provide substantive protections for specific aspects of the environment; NEPA is about procedure. It does not mandate one outcome over another. Instead, it requires that information be gathered, made public, and interpreted. In other words, NEPA is a framework meant to require agencies to disclose information. It is not a substantive law protecting the environment.

Under the stipulations of NEPA, before a project can begin on federal lands or with federal funds, project developers must write a "detailed statement" on the expected environmental impacts of the proposed action.[8] The Council on Environmental Quality (CEQ), as established by NEPA, is responsible for making sure the act is implemented correctly. The CEQ has defined *detailed statement* to mean either an environmental assessment or an EIS.[9] An EIS, the more analytic and lengthy of the two NEPA documents, is required if "significant impacts" are expected from a project. An assessment is generally less intensive and is required when a project is not expected to have significant impacts, or if it is not known what the expected impacts may be. Categorical exclusions are even simpler documents and are granted for routine agency actions with no expected environmental impacts. Federal agencies use these three documents to assess the overall impacts of a proposed project on federal lands and make these documents available to the public.

NEPA also calls for analysis of the status quo by requiring that the project be compared to a "baseline" or "no-action" alternative. In an EIS, "reasonable alternatives" to the proposed action are rigorously explored and objectively evaluated.[10] These potential impacts are then published in draft form, released to the public and agencies for comment, revised based on received comments, and finalized. After the NEPA document is finalized, a record of decision is issued for an EIS or a finding of no significant impact is issued for an environmental assessment.

Over the years, the courts have consistently held that NEPA is primarily a procedural law that has no teeth for actual enforcement.[11] Thus, the focus within NEPA documents is to follow procedure "to the letter" because vulnerabilities arise not from matters of enforcement but from potential litigation or claims that procedures have not been followed properly. Choosing an environmentally friendly alternative in the NEPA process is not required. In fact, "for all of its extensive analysis and public disclosure requirements, NEPA demands no results."[12]

In the years since President Nixon signed the law, NEPA has taken on a life of its own. Every federal agency is required to comply with NEPA, as is the private sector for projects that involve any federal lands or funds. Every time the government does almost anything, someone must be there to write a NEPA document. This means NEPA can cause real delays and have significant costs for many projects.

According to the CEQ, a range of 45,000 to 50,000 environmental assessments and a range of 350 to 450 EISs are prepared every year.[13] The U.S. Department of Energy (DOE) estimated that it takes, on average, thirteen months for them to complete an assessment, at an average cost of $110,000 each.[14] For the more complex EIS, it took the DOE, on average, thirty-three months to complete the report, at an average cost of $5.8 million each.[15] The national average completion time for an EIS is actually even higher, at forty months.[16]

Using the DOE ten-year average and numeric data from the CEQ, we estimate that the federal government spends over $5.5 billion on NEPA environmental assessments and over $2 billion on EISs every year. This means the federal government is annually spending $7.5 billion on fulfilling just a portion of NEPA. For the DOE, only 2 percent of the agency's actions consist of preparing EAs and EISs. The other 98 percent are categorical exclusions, whose costs are not tracked.[17] Even more alarming, this number also does not include costs to private individuals who must comply with NEPA, the cost of federal employees doing NEPA work on non-agency projects, the opportunity costs of time delays or lost investments, or litigation against NEPA documents that have been finalized.

This back-of-the-envelope estimate of the costs of NEPA is just the tip of the iceberg. In the last forty-four years, no government study has sought to tackle a comprehensive quantification of the costs associated with NEPA compliance. In a request for the Government Accounting Office to do just that, Public Lands and Environmental Regulation Subcommittee Chairman Rob Bishop of Utah noted, "the full extent of its (NEPA's) breadth and reach, spanning across every federal agency and every federal action, has yet to be examined. Nor have the costs associated with compliance been fully measured."[18] At this point, we can only imagine the billions of taxpayer dollars expended through the NEPA process.

Many energy projects in particular face permitting challenges, and NEPA is often a substantial part of those challenges. A study by the U.S. Chamber of Commerce determined that the nation's permitting process, including NEPA, is holding up $3.4 trillion in GDP and 1 million jobs per year.[19] It is difficult to say exactly what portion of those trillions of dollars and millions of jobs are not being realized due to NEPA, but it is clearly significant.

NEPA and the Balance of Nature

Like many of the other laws discussed in this book, NEPA is the result of misguided 1960s environmental thought known as the balance of nature ideology. This belief holds that nature is static and predictable and that it can be returned to an inherent balance. The NEPA's simplistic approach is true to the era in which it was written and fails to take into account the complex nature of ecosystems.[20] As Daniel R. Mandelker describes it, NEPA relies on the assumption that bureaucrats are able to rationally predict environmental change and act accordingly.[21] Unfortunately, "this hope disappeared with the understanding that environmental systems are complex, dynamic, nonlinear, and mutually independent."[22] Because nature is always in flux, predicting environmental impacts as required by NEPA is much more difficult than the crafters of the law ever imagined.

While NEPA does not speak directly of a balance of nature, the concept is implicit in its requirements. For example, requiring that project actions be compared to a baseline assumes that a baseline actually exists, and that any human action would disrupt this balanced, ideal state. Any potential action must be compared to the alternative of not doing anything.[23] It should not be surprising that environmental interests have historically favored the no-action alternative and have used this part of NEPA to obstruct projects in favor of an environment they view as better left alone.

Because the balance of nature ideology is no longer recognized as correct, and because NEPA fails to meet its statutory objectives, many are calling for NEPA to be revised or even rescinded.[24] Although very few changes have been made to NEPA in last forty-four years, there have been substantial changes in thought regarding the balance of nature. In fact, the theory is held in contempt by most contemporary ecologists.[25] Mandelker notes that there are fundamental "issues in NEPA's statutory and regulatory structure that demand attention . . . in light of this newer perspective on environmental management."[26]

NEPA as a Tool for Political Entrepreneurs

Although most ecologists have rejected the balance of nature ideology, one reason it is still in play is that it can easily be used as a tool for interest groups to get what they want. Like the Endangered Species Act, NEPA has become a tool wielded by environmental groups, agency personnel, and members of the public who oppose projects close to their homes. These political entrepreneurs are often called NIMBYs (Not in My Backyard) not because they oppose projects for ideological or political reasons, but simply because the projects are too close to an area they value highly. The NEPA already causes significant costs. Those costs are increased exponentially when political entrepreneurs manipulate the law, and although taxpayers see no benefits from these obstructionist tactics, they are the ones who end up paying the bill.

The incentive structure of NEPA encourages project opponents to use whatever means possible to stall projects. This often includes maximizing costs and delays by tying projects up in litigation. As Jim Vines et al. noted, "NEPA review and related litigation tie up billions of dollars of investment and slow job growth by delaying energy projects or making them uneconomical to complete altogether."[27] In other words, political entrepreneurs can get what they want using obstructionist tactics like litigation. One radical environmentalist group called Enviro Defenders publishes a "Legal Handbook for Environmental Activists" on its website. The handbook instructs activists on how to use NEPA to increase costs, push agencies, delay projects, and ultimately sue those pursuing the project.[28]

Creating costs and generating obstructionist tactics is not what legislators intended when NEPA was written. "Lawmakers appear to have been relatively naïve about the limits of environmental science, the machinations of bureaucratic self-interest, and the distortions of policy wrought by judicial activism."[29] The detailed statement required by NEPA has become literally thousands of pages that may require a decade or more to produce. Doc Hastings, chairman of the House Committee on Natural Resources agrees, stating, "After forty years, mounds of red-tape and exposure to endless litigation and regulatory delay have corroded the original intent of the law. The result is a costly and complex regulatory process that does more to impede job creation and economic growth than protect the environment."[30]

Although NEPA was intended to help agencies consider the environment when making decisions, NEPA is not getting that right, either.[31] Environmental groups, such as the American Forest Resource Council, also recognize that NEPA's shortcomings are causing public resources to deteriorate:

> NEPA has evolved into a logjam of overwhelming scale and proportions. [It] is actually causing forest watersheds and habitats to deteriorate as a result of litigation, appeals, and gridlock. Without improvements in NEPA, including modernizing this common sense law and its regulations, I have little hope that our land managers will be able to get back to managing and protecting forests, key watersheds, critical wildlife habitats, rural communities and people.[32]

Thus, political entrepreneurs who want to obstruct projects altogether are getting in the way of political entrepreneurs who use political capital to demand proper management of the environment. Again, taxpayers see no benefits from these conflicts but pay for all of it with higher taxes and poorly managed lands.

The Political Entrepreneur Methodology

The incentive structure created by NEPA is not difficult to understand. It is only rational for political entrepreneurs to use NEPA to do whatever they can to stall a project. Colloquially, this is known as being a "pain in the ass." To make the "pain" go away, project managers will typically bend to the entrepreneur's will. Often, the mere threat of using obstructionist tactics is enough to get projects to change direction or stop altogether. In addition, bureaucratic technicalities and agency workloads can be used to delay and stall projects.

Built into NEPA are tools that allow costs and delays to be maximized by those who wish to subvert projects. The specific ways political entrepreneurs can use NEPA to increase costs and delay project progress are myriad. Some are more overt than others, and some are simply rational responses to incentives created by the law and its bureaucracy. Regardless, these perverse methodologies have real costs to taxpayers and will be revealed and discussed below. These examples of NEPA-related tactics used by political entrepreneurs to delay, obstruct, and profit from projects are taken both from the direct experience of the authors and from applicable literature.

Delay Tactics

As the time to complete a project increases, so does its cost. Therefore, delaying a project is an effective tactic. If political entrepreneurs can increase project costs, they become more pressing for project managers and drive the latter an inch closer to exhausting project budgets, pushing the project toward bankruptcy. Although true bankruptcy with federally funded projects is rare, incremental cost escalations can influence project managers to bend to the will of obstructionists. As we will see, NEPA offers a variety of ways to be an obstructionist throughout the project timeframe.

Detailed Analysis

As agencies and environmental groups review public NEPA documents prior to finalization, a frequent tactic is to continually call for "more detailed analysis" of a given resource. Use of this tactic is well documented and is known as "paralysis by analysis."[33]

Although NEPA case law precedent requires that a NEPA document demonstrate only a "hard look" at affected resources, most NEPA project managers will agree that a more detailed look at a resource can be taken.[34] After all, NEPA managers from one agency often work closely with NEPA personnel from other agencies, and doing a little more work to satisfy a colleague is not too much to ask. In fact, the responding NEPA specialist may be asking the requester to do the same thing on a project in the near future. In this way, asking for more analysis can be part of a tit-for-tat strategy in which agencies cooperate.

Often, this detailed analysis adds little to the actual disclosure of impacts, and it may verge on speculation. A report from the CEQ acknowledged that impact analyses are largely based on assumptions with weak predictive quality.[35] More importantly, such analysis gives decision makers scant relevant information on which to base decisions. These routine requests for additional analysis in NEPA documents can add up to substantial time and expense for project managers and taxpayers, without improving the quality of the document and, according to some, can even be counterproductive.[36]

In our experience, agency and organization personnel will often cite numerous potential, if improbable, impacts to a resource and then demand a full-blown detailed analysis. Often, these requests from agencies come not from having data about the resource but from not having any data at all. In one project, U.S. Fish and Wildlife Service personnel noted that a project in the vicinity of a wetland area could impact dozens of species of birds. As a result, the agency requested a multi-year study as part of the EIS on the impacts to each species for a variety of factors. That study alone would total over $300,000 and reveal little more than the fact that the project could impact dozens of species of birds. Little additional accuracy or precision regarding impacts was gained through the study, and certainly no information that would add to decision-making effectiveness, but the project was significantly delayed and costs increased.

Over time, NEPA managers have come to realize that requests such as these are part of the game. Rather than wait for resource agencies to request certain studies, these studies are simply planned from the beginning. Despite clear guidance about the intended length of NEPA documents in the law (150 to 300 pages for EISs and 10 to 15 pages for environmental assessments), requests for more detailed analysis routinely drive the page count of assessments into the hundreds and EISs to more occasionally several thousands. One author calls the EIS process *documentum infinitum*, as "unimportant aspects of the proposal are documented in excruciating detail."[37] Many EISs are so large that they must be printed in multiple volumes.

Once a certain level of detail is included in one NEPA document, it is likely to reappear duplicated or exceeded in the next document. After all, no Fish and Wildlife Service biologist wants to see only a single paragraph for the lesser prairie chicken, and no soil scientist with Natural Resource Conservation Service wants to review an EIS without the expected write-up on Bashaw Silty Clay Loam microorganisms. In short, as particular "ologists" review NEPA documents, they fully expect that their "ology" will receive as much attention as it did in previous NEPA documents. These expectations help drive up the length and cost of NEPA documents.

Researchers Susan M. Smillie and Lucinda Low Swartz also blame the EPA's Office of General Counsel for "excessively long" documents.[38] They claim that agency lawyers "demand that a great deal of frequently useless information be included in a NEPA document" in an effort to avoid litigation."[39]

There is no real accountability—or limitation—for those requesting more detailed information. Both agencies and public parties are free to speculate and request over and over that the project address those speculations in the NEPA document. As Smillie and Swartz point out, "anyone can claim that a NEPA document is inadequate."[40] As environmental groups review NEPA documents at the draft stage, comments on each section of the document, regardless of the level of analysis included, all tend to read the same: "more analysis needed."

Thomas Margro, who at the time of this writing has been working for ten years to get a toll-road project in California through the NEPA process, testified in 2005 before Congress in support of modifying NEPA. He explains,

> Resource agencies feel unconstrained in raising issues or requesting studies on a piecemeal basis often without considering whether the issues were already addressed or whether the agency requesting the information has any rational basis for doing so.[41]

Indeed, an agency's first assumption in a NEPA project is that additional information is needed, paving the way for analysis paralysis.

On the ground, it is very difficult for project managers to say no to requests for more analysis from political entrepreneurs. Often, you are sitting across the table from a colleague who just told you that you are giving insufficient attention to the resource to which he or she has devoted his or her life. You may also be sitting across the table from an environmental activist who says that if you do not properly analyze the possible changes to the color of a frog's toes that might result from your project, you will next hear from him on the 10 o'clock news. As a NEPA manager, the last thing you want is to generate more bad press for your agency or to anger a friend, so you are inclined to give in to both the requests. It's no wonder a congressional task force in 2005 found that the costs of complying with NEPA are rising.[42]

Delay Real Concerns or Issues

As NEPA projects get off the ground, there is a dedicated period called *scoping*. During this period, information is gathered regarding issues that will likely need to be addressed in the NEPA document. These issues may include wildlife, wetlands, or transportation, and the process includes coordination with both the public and agencies. Savvy political entrepreneurs, however, have no incentive to divulge issues or concerns up-front—not when they can do so later and possibly increase time and costs.

Prior to finalization of NEPA documents, a draft document is released for review by the public and government agencies. By this stage of the NEPA process, personnel have invested substantially in preparing a draft document. Research has been collected, analyses completed, studies and assessments have been finalized, and results have been recorded and integrated into the document. Political entrepreneurs know that substantial changes to a draft NEPA document can have dramatic effects, and so they will frequently delay disclosure of potential issues until this point. New issues revealed this late in the process can cause substantial delays and may require analyses to be redone or new analyses to be initiated. In the case of EISs, it can often take several years to get to this point in the NEPA process while studies and assessments move along. If a new study is required after a draft document is complete, delays can easily add as much as several additional years.

The June Sucker Fish

The case of the June sucker is one example of such delay during the preparation of NEPA documentation. The June sucker is a fish native only to Utah Lake and Provo River in Utah. It is protected under the Endangered Species Act, and multiple efforts are under way to help recover the species. The June sucker spawns and lays eggs in the Provo River in June (hence the name). Historically, the eggs floated down the river to Utah Lake, where small fish would hatch and grow in the protection of marshy areas along the shoreline. Once the fish were big enough to avoid predation, they would swim into the open waters of Utah Lake and then eventually head back to the Provo River to spawn.

Vast changes to Utah Lake and the Provo River over the last century have endangered the fish. The river has been channelized to reduce flooding, the marshy refuge areas are now farmland, the channel has been regularly dredged to deepen it, the lake is full of exotic species of fish and plants, and water flows now come from the bottom of an upstream dam, changing the temperature of the water and the timing of the flows. As a result of these changes, not a single juvenile June sucker had been documented in the last forty years. It is suspected that most June sucker eggs are eaten by exotic fish in the deep, slow-moving water of the channelized portion of the river. The lake has also been invaded by phragmites, an invasive aquatic weed that chokes out native vegetation and offers small fish no protection. In the early 1990s, prior to the construction of a June sucker fish hatchery, it was estimated that fewer than 500 June suckers survived. At that time, portions of the Provo River were designated as "critical habitat" under the Endangered Species Act, and the June sucker was listed as an endangered species.

In 2007 a Federal Highway Administration project proposed that a road be built to the Provo Airport from U.S. Interstate 15, a distance of just over three miles. Part of the roadway would be built in the historic floodplain of Utah Lake, an area that consisted of residential backyards and farm fields. The project produced a draft EIS document that covered the plight of the June sucker and the floodplain in detail. Upon review of the draft EIS, a Fish and Wildlife Service employee, a vocal project opponent from the start, suddenly remembered that if a variety of unlikely events occurred simultaneously, the project could affect the June sucker.

Although the project area was miles from the edge of Utah Lake, if the lake were to flood sufficiently, the new roadway would be surrounded by water, the employee said. The last time this happened in the area was thirty years earlier, before changes to the Utah Lake dam structure made potential flooding even less likely. Despite the unlikely nature of the events, these potential impacts to the floodplain itself were disclosed in the draft EIS.

Not disclosed, however, was the even more remote scenario under which the proposed road's presence could affect June sucker "rearing" habitat for juvenile fish. This scenario relied on another variety of unlikely simultaneous occurrences that could mean the proposed road would replace what could be endangered species habitat.

So, what are the events that would have to happen simultaneously in order to have a "potential impact" to June sucker? First, the lake would have to flood to a certain height, something it had not done in thirty years and that may not have even been possible any longer, given changes to the dam on Utah Lake. Then, the lake would have to flood for a duration sufficient for the right species of vegetation to sprout underwater and grow to a height that would protect juvenile fish from predators. This vegetation would need to sprout in areas that were at the time residential yards, city roads, and farm fields. Then, the June sucker would have to be able to successfully reproduce, which had not happened in forty years. Finally, the juvenile fish would actually have to get to the area where the project intersected the floodplain, a distance of about four miles, through predator-infested waters with little to no cover for small fish. The Fish and Wildlife Service's revelation suggested that the June sucker would be impacted only after a dizzyingly unlikely series of circumstances.

Even if all of these events were to happen simultaneously, "impacts" were still not guaranteed. In fact, if all of these things were to happen simultaneously in the project area, they would have also occurred in many other places on Utah Lake. In such a case, the small exclusion of the project area (or road footprint) was likely to be so small as to not matter at all to the survival of the June sucker.

Despite the unlikeliness of these events, the Fish and Wildlife Service, following the draft EIS, required a formal biological assessment, formal consultation under Section 7 of the Endangered Species Act, and additional analysis and documentation in the EIS. This process singlehandedly added one year to the project, cost hundreds of thousands of dollars, and could have been completed concurrent to other EIS resource assessments if the issue had been identified at the beginning of the project. More appropriately, the Fish and Wildlife Service could have admitted the ridiculousness of the impact scenario and dismissed the need for the analysis.

As this example demonstrates, political entrepreneurs have learned to obstruct projects by withholding scoping issues until late in the NEPA process. This tactic, made possible by NEPA itself, will continue to be used to delay projects and increase project costs because it has direct benefits to political entrepreneurs.

Insufficient Alternatives Analysis

Both the public and agencies participating in NEPA analysis often use "insufficient alternatives analysis" as a tool to delay or stall a project. Despite requirements within NEPA that dictate documents are to be "brief" and "concise," political entrepreneurs will request that a particular alternative to the project be analyzed in detail. Similar to delaying issues identification, if new alternatives are identified late in the process, then delays and costs can be maximized. J. Matthew Haws points out that the "'reasonable alternatives' requirement can be used to frustrate the objectives of the proponent, the cooperating agencies, and local interests alike."[43]

Recent White House guidance on NEPA reiterates original regulation mandates for streamlined documentation.[44] Despite this order, agencies and political entrepreneurs continue to demand greater and greater detail and information regarding both the alternatives analyzed within the document and the alternatives that are dismissed from further consideration. This additional and unneeded analysis creates costs and delays for projects during the NEPA process.

In a NEPA document, a variety of alternatives to the project—if there are any—are analyzed, including a no-action alternative. For example, in the case of a transportation project, the purpose may be to get people from point A to point B. The NEPA document would likely analyze the proposed project, several alternatives to the proposed project, and a do-nothing alternative (no-action). Obviously, there are numerous ways to get from point A to point B: by car via route 1, by car via route 3, by bus, by hovercraft, by airplane, by foot, and so on.

As part of the NEPA process, project personnel, usually with input from the public and agencies, whittle down the alternatives to those that are reasonable, given the project context and other factors. Reasonable alternatives are then analyzed in detail in the NEPA document at great expense. Because of this, not every single alternative to a project is analyzed in great detail, and some are excluded from any detailed analysis.

Attempts to force analysis of every minute variation of alternatives will inevitably cost more money and take more time; they actually undermine the intent of NEPA. Instead of presenting well-researched analysis on potential environmental impacts, decision makers get "a superficial consideration of a multitude of remote alternatives."[45] The draft NEPA document provides yet another opportunity for political entrepreneurs like agencies, organizations, and the public to come up with previously un-analyzed alternatives. As with other tactics, discovery of a new alternative at this time in the process, and its subsequent analysis, can have dramatic effects on project timeframes and budgets.

Political entrepreneurs from the U.S. Army Corps of Engineers used this tactic on a recent project by suddenly claiming that a reasonable alternative had been dismissed from detailed analysis prematurely. The potential alternative to a new road in a rural area consisted of an "elevated expressway" directly over an existing rural road (similar to elevated roads in New York City or Chicago, but in a rural area). Basically, the Corps of Engineers asked, why not build a road above an existing road? The ridiculousness of this alternative is obvious, but it should be noted that "reasonableness under NEPA is subjective, and political entrepreneurs will use whatever tools are at their disposal. Because the Corps of Engineers also wields Section 404 of the Clean Water Act (see Chapter 7 for more on this topic), project management agreed to analyze its proposed alternative—at great expense. In this case, detailed analysis of this alternative took an additional six months and cost tens of thousands of dollars.

Agency personnel who insisted on the analysis of the additional alternative had been participating with the project from the beginning and were aware of the alternative's previous dismissal. As noted, however, agency personnel have zero accountability and no incentive to meet project deadlines or save taxpayer money. By using obstructionist tactics and waiting to reveal problems with an alternative, political entrepreneurs effectively moved this project to the backburner for six months, increased the amount of time they spent on projects (justifying their jobs and relevance), increased project costs, and delayed implementation of a project they considered unfavorable.

Pet Projects Disguised as Mitigation

Resource agencies, regulatory agencies, and environmental organizations are notorious for their lack of resources and small budgets. Many political entrepreneurs are perpetually poor in terms of funding. Despite these shortfalls, agencies and organizations have long lists of pet projects that they would like to complete in order to justify their existence and please their membership. The Fish and Wildlife Service would like to create more habitat, the U.S. Forest Service would like to develop more recreational facilities, the Sierra Club would like to protect more wetlands, and so on.

As NEPA projects develop, project managers and policymakers will often include enhancements for their pet projects under the guise of mitigation, or preventing negative harm to resources. Merriam-Webster defines mitigation as "to make less severe or painful." In general, mitigation for resource impacts deals directly with that resource: noise impacts are mitigated with noise walls, impacts to birds are mitigated by enhancing bird habitat, impacts to farmers are mitigated with more farmland, and so on. Mitigation under NEPA is often more of a buy-off. Studies of particular resources, purchases of land for preservation, and other "scratch my back and I'll scratch yours" requests are viewed by some as an exchange for other favorable outcomes from that agency.

When the City of Afton, Wyoming, initiated the NEPA process to reinstall a hydroelectric plant that was destroyed by an avalanche decades earlier, the Forest Service jumped on the opportunity to get more data on possible Canada lynx habitat. They accomplished this pet project by requiring extensive and expensive surveys despite the fact that the project was not in designated critical habitat for Canada lynx and was using existing infrastructure and not affecting any previously undisturbed land. What's more, the Forest Service then required the project to build a parking lot at a nearby recreational area to mitigate possible impacts to Canada lynx habitat.[46] How exactly does a parking lot make impacts to Canada lynx "less severe or painful"?

By using their authority in the NEPA process and demanding mitigation, the political entrepreneurs at the Forest Service gained valuable data and saved money by not having to fund their own study or parking lot. This misuse of mitigation and authority under NEPA incrementally delays projects and increases their costs.

Not only is mitigation being demanded in greater and greater quantities for unrelated impacts, but the definition of mitigation is being lost along the way. Mitigation is not supposed to be used to enhance recreation areas, as in the example above; it is simply meant to ease potential harm associated with change. In truth, Richard Fristik says, "NEPA does not pose a substantive requirement upon agencies to mitigate."[47] Let's remember, however, that NEPA managers must work closely with colleagues from other agencies and that "what you ask for today I'll be asking for tomorrow." Some agency employees have only the concessions given in the NEPA process to demonstrate usefulness. That is, their value is partially determined by how much mitigation they can get during the NEPA process.

Fristik notes, "mitigation can become a catch-all of sorts, including any and all measures thought necessary to ameliorate project effects, be they real or perceived."[48] Indeed, our experience is rife with examples of EISs that demonstrate no impacts to resources, but wherein agencies or organizations propose mitigation anyway. There is the feeling that any project is inherently bad and surely affects something. Fristik shares the frustration of this misuse of mitigation by stating, "Seemingly, if there is no 'mitigation' for every possible impact, whether genuine or not, the analysis is somehow considered incomplete or incorrect."[49]

When mitigation becomes a buy-off rather than an action taken to ease the pain of an impact, bad things can happen. The pitfalls of mitigation in the NEPA process are evident in the case of the Ruby Pipeline, a natural gas pipeline between Wyoming and Oregon proposed by the El Paso Corporation. In an effort to stave off a suit by the Western Watersheds Project, the El Paso Corporation reached a deal to set up and fund two conservation trusts to the tune of $22 million. This form of mitigation actually may have backfired, as the Center for Biological Diversity brought the same suit. Although El Paso had bought off one organization, the suit was simply filed by another sister organization.

Overstating Likely Impacts

Nothing works like exaggeration and misinformation to get people fired up about the environment. Because NEPA allows for public and agency participation, there is a built-in mechanism whereby political entrepreneurs can derail and subvert the project in the court of public opinion. Time and again, the public works itself into a frenzy about the supposed results of a particular project. Agency personnel, who see themselves as resource guardians, are not immune to these tirades. To justify his inaccurate speculation about likely impacts of a recent project, a chief regulator with the Corps of Engineers once told us that he felt he had a moral obligation to protect wetlands in the "arid West" beyond what is required by law because they are more rare than in the East. This paternalistic protection of resources by agency personnel, often beyond what is required by law, can be the source of some serious project opposition that leads to greater delay and cost overruns.

Nearly every NEPA project involves public speculation and agency personnel overstating likely impacts as a project gets off the ground. Individuals opposed to the project often work with environmental organizations to speculate on how projects could potentially cause serious and irreparable harm. In addition, environmental groups lobby agency personnel to convince them of the detrimental aspects. In this way, environmental groups can use agency personnel and expertise to accomplish what they themselves may not be able to do.

Savvy political entrepreneurs opposed to NEPA projects have become adept at getting their speculation about what they see as potential impacts into news media outlets prior to the NEPA analysis being completed. In skilled hands, false information can modify or halt projects. As misinformation takes root, pressure can build on project managers to complete additional studies, change the project goals, slow timeframes, or stop a project altogether.

During the NEPA process of a proposed river restoration project in Utah, a project opponent misconstrued conversations with project personnel, exaggerated impacts, and speculated on what would happen as a result of the project before the NEPA process was even done. Newspapers printed his fabrications, and websites published interviews (see, for example, reporter Caleb Warnick's January 12, 2012 article in the Provo, Utah *Daily Herald* about the project). The project opponent claimed the project would shut down the river and kill trees, even though the project would actually increase access to the river, benefit existing trees, and plant over a hundred more acres of trees.

This bad press, coupled with inaccurate information about impacts, created intense public pressure and caused the project to conduct three additional studies for the NEPA document. These studies, and the additional public outreach, delayed the project a full year simply to put public misconceptions to rest. The project manager, to satisfy the misinformed public, was basically forced to publicly "take a step back" to demonstrate a fair and inclusive process so the project could eventually move forward.

Backward steps are costly, but political entrepreneurs who oppose a project do not have to pay those costs, which are paid by project and agency personnel who have to get their procedures right or be sued. Exaggerating the impacts and potential outcomes allows project opponents to gain ground as they increase costs and delay the project.

We are not saying that federal projects are without fault. Many of them make no economic or ecological sense. We are just pointing out how NEPA can be used to stall, delay, or kill projects without anyone being responsible for determining whether the project makes sense in the first place.

Controversy

Under NEPA, it is appropriate to analyze many projects with a categorical exclusion or environmental assessment. If, however, a project opponent is able to demonstrate "significant controversy," then NEPA and agency implementation regulations dictate that the project must be analyzed as an EIS.[50] Recall the average times to complete an environmental assessment versus an EIS: about a year versus 3.5 years. Recall also the average cost for an assessment was $110,000, and the average EIS cost was $5.8 million. Thus, political entrepreneurs have an incentive to create controversy from the beginning of a project because it may cost the project more time and money.

William Murray Tabb notes the incentives attached to NEPA regarding controversy:

> Agencies actually have a marked disincentive to pursue public involvement. This is because the practical effect of controversy, if it rises to a certain level, is that the agency may be required to undertake additional duties of study and analysis.[51]

The better question is why would political entrepreneurs *not* stir up controversy? The benefits to their cause could be huge.

Project managers are primarily concerned with and focused on their projects and their outcomes. As a result, when the Fish and Wildlife Service representative for their project and the EPA project liaison both say that controversy surrounding the project is high, project managers consider writing an EIS. After all, managers may need to rely on the Fish and Wildlife Service to help with a biological opinion later in the project, and they definitely need to have their document reviewed by EPA, so it's better not to get on the bad side of potential opponents.

As controversy takes over a project, NEPA requires a higher degree of analysis with the subsequent increased costs in terms of time and money. With regard to categorical exclusion and environmental assessment projects, no obstructionist tactic has more potential than to elevate the controversy.

Insufficient Resources / Additional Time

Agency bureaucrats deal with federal law, executive orders, agency guidance, agency policies, and many other types of regulation that rule the majority of the workweek. As such, many agency employees become adept at stonewalling techniques that they routinely employ when a NEPA project is taking up too much time or is deemed unfavorable. But the use of bureaucratic tactics is not limited to government agencies. Often, the same obstructionist techniques that work for agency personnel also work for project opponents that belong to organizations.

Federal agencies tasked with conducting NEPA studies suffer from small budgets and insufficient staff.[52] Many of the agencies tasked with conducting NEPA analyses, particularly those associated with public lands, such as the Forest Service and BLM, have insufficient resources to conduct an analysis within reasonable timeframes. As a result, private interests must pay for the analysis themselves through expensive consultants or wait as long as five to ten years for the agency itself to conduct the analysis. Waiting or paying for analysis of even the simplest projects has impacts on project resources and viability. The seriousness of the agency funding problem is evident in the statement of Ms. Dinah Bear, who served for twenty-five years as general counsel to the

Council on Environmental Quality: "The environmental review process itself is much less of an impediment to permitting and construction than the lack of adequate staffing and resources at Federal agencies."[53]

In a recent project, the authors worked with a man who owned a private inholding on BLM land. The land had been in his family for generations and was located in an extremely remote area forty miles from the nearest town. The gravel road accessing this three-acre inholding had, over the years, deteriorated. Trees and vegetation adjacent to the road had encroached, making it difficult to access the property. The property owner requested that the BLM allow him to regrade the single-lane 1.3-mile road, apply new gravel, and trim back the vegetation about three feet on each side of the road. The small inholding was located in the middle of thousands of acres of pinion-juniper forest and was used only for five or six months of the year.

The BLM, citing NEPA, required that an environmental assessment be prepared to determine if significant impacts could be expected to the environment as a consequence of this meager road improvement. The BLM office could do the assessment, but this would take three to five years because of its insufficient resources. The other option, they told the landowner, was to pay a consultant to start the process immediately. Given the road's current condition, the landowner could not afford to wait three to five years and opted to have a consultant prepare the environmental assessment for him. Preparation cost the landowner $15,000 and took nearly nine months. Following the finalization of the process, the landowner spent about $5,000 and completed what was essentially a driveway repair. Insufficient federal agency resources for even the smallest projects helped quadruple the landowner's cost and delay the project for an entire season of use.

Agencies also use claims about insufficient resources during the NEPA review process, which requires reviews and comments from the agencies with jurisdiction over particular resources: Fish and Wildlife Service for wildlife, EPA for air quality, and so on. As requests for comments are met with increased delays, the same refrain is often heard: "We don't have enough people to get to that right now." There is simply no incentive for an agency to prioritize another NEPA project above internal projects. Thus, timelines, where they exist, are maximized, and requests for additional review times are common. Even worse, some agencies claim to have no comments when they have not actually taken the time to review, only to discover issues when they actually get to it at a later, and more inconvenient, time.

The authors participated in a controversial NEPA project in the western United States, in which EPA personnel suggested the use of a professional moderator to reopen channels of communication between agencies and help find common ground. The project, at great expense, hired a professional, nonaffiliated, unbiased moderator and then set up a meeting between all the agencies involved. Schedules were double-checked and commitments to attend were finalized from all parties. Substantial time and monetary investments were made to secure facilities and food for the day-long meeting. Several representatives from the project sponsor flew out from Washington, D.C.

Even though the moderated meeting was their idea, the EPA, a vocal opponent of the project, called the day before the meeting to inform project personnel "there is no money left in the budget for us to fly out from Denver and attend a meeting." Even the local Fish and Wildlife representative decided to join the meeting by phone, citing "insufficient resources and a busy workload." The meeting, without key people actually in attendance, was a flop and a waste of time and money.

Although logistical and planning problems are common, so too are the last-second claims of insufficient resources and requests for additional time to review NEPA documents by political entrepreneurs. Although these bureaucratic tactics to delay NEPA projects are not generally significant contributors to time delays and budget overruns, they cumulatively and incrementally increase costs and delays to NEPA projects.

Cover Your Ass and Litigation

The incentive to "cover your ass" is well documented among agency employees. It involves ensuring that procedures and regulations have been followed so that you cannot possibly be responsible when something goes wrong. Because procedure is the heart of NEPA, accomplishing this can take on epic proportions as NEPA personnel attempt to create "litigation-proof" environmental assessments and EISs. Implicit in this mythical NEPA document is the idea that every resource is analyzed in minute detail and every NEPA procedure, guideline, and nuance has been followed, documented, and rubbed in the face of would-be litigants.

This culture stems from the fear of agency employees that their projects may be challenged in court. Although only a small percentage of projects are actually subject to lawsuit, the mere idea or threat is enough to make agency officials spend whatever is necessary to have a bulletproof NEPA document. To do this, agencies must spend a lot of money, employ many people, and take their time. Haws notes that "litigation, or the threat thereof, is the most powerful enforcement mechanism in the NEPA process."[54]

Consultants love the pursuit of the litigation-proof EIS. Maximum time and money are spent on each resource, ensuring that everything that could be said about it in the analysis is actually included. Coordination times with agency personnel are maximized in an effort to get buy-off on resource analysis prior to the draft document. Quality control and quality assurance costs skyrocket as senior personnel review documents repeatedly. Often, attorneys are consulted to ensure NEPA compliance on every minute point.

A project we participated in recently employed a specialized NEPA attorney at $535 per hour. A half-hour phone call cost nearly $300, and two weeks of work cost over $42,000. This was a small project in the NEPA world, so one can imagine the attorney fees on a big project involving multiple attorneys for the duration of the project.

It is no wonder project personnel are litigation-shy. "Courts decide cases relating to NEPA more than any other environmental statute," and agencies must "operate in the shadow of court challenges."[55] Delays resulting from litigation can add years and even decades to projects. The incentives of NEPA encourage litigation. It is important to emphasize that taxpayers are the real losers in this game. Not only do they pay for increased costs, but as Haws pointed out above, they also miss out on projects that may benefit them once completed.

Vines et al. point out that an average of 126 NEPA cases were the subject of lawsuits every year between 2001 and 2009.[56] The same time period also saw an average of twenty-four preliminary and permanent injunctions each year.[57] These numbers may appear small in the context of the national preparation of over 50,000 NEPA documents annually, but it is the threat of litigation, not the litigation itself, that motivates agencies to maximize time and budgets on bulletproof NEPA documents.[58]

In 2005 a lawsuit was filed against the Federal Housing Authority and the Department of Housing and Urban Development "for not completing a NEPA analysis before issuing each and every insurance and loan guarantee."[59] Under the Administrative Procedure Act, plaintiffs have six years following a final agency action to initiate NEPA litigation. Given the near certainty that its NEPA documents would be litigated, Shell Oil Company filed a lawsuit *challenging its own project* in order to avoid waiting the six years before the statute of limitations expired.[60]

The increase in NEPA litigation and the resultant delays are also having environmental impacts themselves, helping to negate any environmental benefit from NEPA. Haws notes that lawsuits "are often guided by interest groups seeking immediate benefits for themselves rather than the public benefits of such action."[61] Agency personnel are also noticing the negative impacts of NEPA delays. Michael Coulter quotes the regional forester for the Forest Service in Washington State from a NEPA task force hearing, saying:

> We have 44 projects in some stage of litigation right now. Each time we go through the appeal process or the courts, much of our limited resources are employed to defend the decisions we feel are crucial to restoring ecosystems and addressing forest health concerns. Delays in restoration and forest health treatments compound the problem as more acres move into conditions that promote invasions of exotics, leave forests susceptible to insect and disease, and predispose ecosystems to unwanted wildfire.[62]

The quote above illustrates how NEPA's inherent costs and delays draw resources away from responsible land management, increasing the waste of taxpayer dollars. In the early days of NEPA implementation, Fran Hoffinger foresaw the negative aspects of today's rampant NEPA litigation:

> Once the litigation process is undertaken by the private sector, it may necessarily delay an otherwise worthwhile project. There also is the danger that public funds will be misused in massive litigation and that the project costs will rise because of the delay in obtaining agency

compliance. Both time and money would, at the very minimum, be inefficiently utilized, if not wasted.[63]

Agency-Specific NEPA Interpretations

When Congress enacted NEPA, they allowed the agencies to come up with their own NEPA implementation guidelines. As such, each agency ended up with a slightly different way of implementing NEPA analyses. As a result, minor quibbles erupt each time agencies work together on NEPA projects or as they review each other's documents. Linda Luther notes that some members of Congress have expressed concerns that "project delays are the result of inefficient interagency coordination."[64] Our experience backs this up. Different agencies can have vastly different NEPA implementation guidelines. Indeed, interpretations of NEPA terms and concepts such as "purpose and need," "reasonable alternatives," and "significant impacts" can vary greatly from one agency to another.

Other NEPA differences can cause delays as agencies sort out their differences. This is particularly acute when a NEPA document from one agency must be adopted by another, such as during the Clean Water Act Section 404 permitting process. Alternative development and the rationale for dismissing alternatives may be so fundamentally different between agencies that work may need to be redone, incurring additional time delays and costs. This scenario played out in one of our projects and created delays of three to four months as differences were worked out.

In 2005 a task force on NEPA identified "clarifying alternatives analysis" as an area of NEPA that may need revisions. Indeed, they admitted, "the rules for screening and evaluating alternatives [are] unclear, and vary greatly from agency to agency."[65] Collectively, and over the course of a project, bureaucratic differences in NEPA implementation between agencies can amount to significant cost increases and time delays.

Obscure Laws and Regulations

Some agencies lack laws with teeth to force substantial changes in NEPA projects. Increasingly, political entrepreneurs turn to obscure laws or creative interpretations of laws to force projects to comply. These interpretations are often the result of agency personnel being lobbied by environmental interests and colluding to oppose projects.

During the BLM driveway project discussed previously, it occurred to a certain BLM biologist that trimming three feet of roadside vegetation could potentially impact possible nesting birds listed under the 1918 Migratory Bird Treaty Act. This law was enacted primarily to prevent the sale of rare bird feathers from the United States to Canada for use in ostentatious ladies hats that resemble an Indian in full headdress. As a result, nearly 800 species of birds are protected. The BLM biologist thought that trimming roadside vegetation on an extended driveway could potentially wound or kill a magpie.

In practice, this law protects nearly every common bird in the United States, including ravens, cardinals, blackbirds, pigeons, and crows. Most of these birds are not now, nor have they ever been in any type of danger, and they are not listed as threated or endangered under the Endangered Species Act. Given that a bird nest can be encountered nearly anytime, anywhere, it's amazing that this law does not stop every project in existence.

Using this 1918 law as a tool to delay such a minor project in the middle of the Great Basin constitutes misuse. Nevertheless, use of this law, in conjunction with NEPA, effectively delayed a straightforward project, increased costs to the property owner four-fold, and also caused construction to be delayed so that it would occur outside the nesting season.

As political entrepreneurs attempt to stall NEPA projects, they can sometimes find the tools they need right in the Code of Federal Regulations. Increasing use of these obscure laws and their creative interpretations will continue to delay projects and suck up project money.

Other Obstructionist Tactics

Increasingly, resource and regulatory agencies are staffed by personnel with conservationist, preservationist, or even obstructionist tendencies. Many view themselves as guardians of resources that can potentially be affected by private projects that are required to go through the NEPA process. As a result, agency personnel use a variety of bureaucratic tactics to stall and stop projects using the provisions of NEPA. Some of these tactics include overstating indirect and cumulative impacts, repeatedly requesting additional information, discounting or ignoring impacts to societies or economies, requiring greater mitigation from those with the ability to pay (such as the U.S. Department of Defense, U.S. Department of Transportation, the oil industry, etc.), changing personnel mid-stream in the project, and reneging on previous commitments. Cumulatively and individually, these tactics cost project proponents and taxpayers money and do little to provide valuable information for decision makers.

Incentives Matter

Despite the roadblocks and obstructions inherent in the NEPA process, some federal agencies, when given the right incentives, have managed to produce NEPA documentation with less time and cost.

In 2005 a NEPA task force was convened by Congress to assess public sentiment regarding NEPA. Environmental groups alleged the task force was on a so-called NEPA "witch hunt." However, among the findings of the task force was that "environmental organizations believed that NEPA has been working fine and needs little change."[66] Five years later, many of those same environmentalists had reversed themselves on NEPA's effectiveness.

In 2009 President Barak Obama unveiled the American Recovery and Reinvestment Act of 2009, also known as the stimulus package. Among its provisions was significant federal funding for "shovel-ready" and "renewable energy" projects. Green energy supporters, seeing their chance to acquire funds to advance the cause of renewable energy, suddenly began to complain about the lengthy and time-consuming NEPA process. Ironically, environmentalists who had once been in favor of using NEPA to delay projects now began to ask if renewable energy projects could somehow be excluded from NEPA or "expedited." Articles popped up arguing that preparing EISs for renewable energy projects should be "accelerated" because these projects are environmentally friendly (Salter, 2011; Rein et al., 2012).[67]

Trevor Salter, who is familiar with the difficulties of the NEPA process, argues that subjecting renewable energy projects to such a process is inconsistent with the country's goal of increasing renewable energy development.

"Moreover, because renewable energy benefits the environment on balance, there is some irony that NEPA . . . significantly delays environmentally beneficial projects."[68]

Altering the NEPA process to favor one type of project over another would be problematic. The federal government would be attempting to pick favorites by assuming that renewable energy projects will have negligible environmental impacts compared to other projects. Unfortunately, the government does not possess the knowledge necessary to successfully pick such favorites, and favoritism can become a form of cronyism.

Agencies have shown themselves capable of completing NEPA documents more quickly, even without a change in the law. For example, the BLM issued an Instruction Memorandum (IM 2011-059), which made wind and solar energy projects a priority. As a result, EISs for solar projects have been completed in as little as nine months.[69] This suggests that part of the problem with NEPA is political. Projects that are seen as "dirty" or not environmentally friendly may be arbitrarily subjected to more strict NEPA review than renewable energy projects.

NEPA documents for oil and gas projects on federal lands continue to stagnate with little chance of being accelerated. This stagnation is occurring despite the fact that oil and gas projects still offer the best option for economic and job growth. As of February 15, 2015, projects that would develop 2,391 wells had been waiting for more than three years to gain approval through the NEPA process. No major projects were approved in 2014. Furthermore, oil and natural gas development is up significantly on private lands since 2008, but has gone down on federal lands.[70]

NEPA is costly in terms of expenditure, delays, and unrealized potential. As the chairman of the Council on Environmental Quality during the George W. Bush administration explained, "We can save the taxpayer millions of dollars if we do this process more effectively."[71] But, NEPA is not about doing things more effectively or efficiently. It is about a process that people have found ways to use for their own, political purposes. There are many entrenched interests with a strong interest in keeping NEPA just as it is. As one of our undergraduate economics professors was fond of saying, "there is no constituency for efficiency."

Conclusions

It is evident that NEPA is broken. Although it can be said that NEPA has had substantive effect by procedurally requiring agencies to consider the environment by preparing NEPA documents, those documents have become meaningless, with little impact on decision-making or on environmental protection. The NEPA process delays the implementation of good and bad ecological and development projects. Furthermore, resources that could be used for the proper management of federal lands are instead used to combat excessive litigation or for mitigation measures that are actually concessions to associates, rivals, or other political entrepreneurs.

NEPA documents are written for the courts and in anticipation of litigation, and few members of the public really have the time, will, or training to care about NEPA analyses. Vines et al. conclude that NEPA litigation is no longer for environmental protection, but to "simply delay the project with the hope of killing it entirely."[72]

We see that NEPA harms the environment rather than protecting it, costs billions of dollars annually, delays beneficial projects, and reduces the ability of agencies to carry out their statutory duties. Based on antiquated environmental theory, NEPA is also a failure because political entrepreneurs have hijacked it.

7

The Clean Water Act

ON JUNE 22, 1969, Ohio's Cuyahoga River burst into flames. While the exact cause of the fire is unknown, an investigation by Cleveland's Bureau of Industrial Wastes asserted that sparks from a passing train might have ignited the river's oil-covered surface.[1] The river fire caught the attention of national media, and *Time* Magazine reported:

> Some river! Chocolate-brown, oily, bubbling with subsurface gases, it oozes rather than flows. "Anyone who falls into the Cuyahoga does not drown," Cleveland's citizens joked grimly. "He decays" . . . It is also—literally—a fire hazard. A few weeks ago, the oil-slicked river burst into flames and burned with such intensity that two railroad bridges spanning it were nearly destroyed.[2]

The ignition of the Cuyahoga River and the public outcry that followed helped to drive the passage of the Clean Water Act of 1972 (CWA).[3] For political entrepreneurs, the fact "that a river could become so polluted to ignite proved that state and local governments . . . were incapable of ensuring adequate levels of environmental protection."[4] Jonathan Adler explains that calls for a top-down approach to conservation were spreading across the nation: "The event transpired as the nation's environmental consciousness was awakening, and searching for symbols of the burgeoning environmental crisis. A river on fire fit that bill."[5]

In addition to the Cuyahoga River fire, several other environmental catastrophes made national headlines at that time, adding to the movement for federal intervention. In 1968 the chemical DDT appeared in 584 of 590 water samples taken across the nation.[6] The following year, a Union Oil Company platform suffered a blowout, spilling 80,000 to 100,000 barrels of oil into the Santa Barbara Channel.[7] That same year, discharges from four food-processing plants killed 26 million fish in Florida.[8]

Cumulatively, these events spurred political entrepreneurs and shifted the conservation focus in the United States from localized and private efforts to massive federal intervention. Ultimately, these headlines led many to falsely believe that adequate local conservation simply wasn't possible.

Clean Water as an Act: The In's and Out's

The CWA, formally known as the Federal Water Pollution Control Act of 1972, was passed as a major show of bipartisanship and widespread support across the country.[9] In fact, the CWA passed unanimously in the U.S. Senate and received only eleven dissenting votes in the U.S. House of Representatives.[10] Established with the goal of eliminating the discharge of all pollutants into the nation's waters by 1985, the CWA has been called the turning point for America's waterways.[11]

In arguing in support of the CWA, entrepreneurs claimed they feared that if environmental protection was left to the states, state governments would "race to the bottom" and "bargain away the long-term health and well-being of their citizens for current economic gain."[12] Evidence, however, does not support this belief and instead demonstrates the success of local and state control.[13]

The fundamental purpose of the CWA was to "restore and maintain the chemical, physical and biological integrity of the Nation's waters."[14] The act prohibited any pollutant discharge into national waterways with limited exceptions. Under the right circumstances, limited pollution is permitted if the polluter complies with the EPA's own specialized system: the National Pollution Discharge Elimination System (NPDES).[15] Permits granted under this system are designed to limit the quantity, type, and concentration of pollutants that a given industry can legally release. The permits also specify specific control technologies for each pollutant as well as compliance deadlines.

The EPA mandates periodic testing in order to monitor these pollutant discharges.[16] Specific requirements are outlined in the NPDES permits that prescribe specific pollution prevention technologies for specific companies. Based on the pollution problems encountered by each individual company, these specifications require comprehensive examinations coupled with periodic testing.[17] In the future, as technological developments advance and enable increased efficiency in pollutant disposal, the NPDES maintains the right to progressively apply stricter restrictions on polluting companies.[18]

National water quality standards are also determined under the provisions of the CWA. The EPA administrator is specifically directed to establish these standards and to create the regulations necessary to achieve the CWA's objectives.[19] Victor B. Flatt argues "the [EPA] Administrator's most important duty is the promulgation of effluent guidelines limiting the discharge of pollutants."[20] Through these efforts to implement top-down antipollution policies, the CWA is primarily aimed at limiting point-source pollutants: those that can be traced to a single discrete source, such as the wastewater outflow of a factory. The CWA does not regulate nonpoint pollutants, which come from various and diffuse sources, such as storm-water runoff from entire cities or agricultural fields. The distinction between point and nonpoint pollution sources will become significant in the following case study regarding Total Maximum Daily Loads (TMDLs).

The EPA offers the following definition for point-source pollutants:

> Any discernible, confined and discrete conveyance, including but not limited to any pipe, ditch, channel, tunnel, conduit, well, discrete fissure, container, rolling stock, concentrated animal feeding operation, or vessel or other floating craft, from which pollutants are or may be discharged. This term does not include agricultural storm water discharges and return flows from irrigated agriculture.[21]

Point-source pollutants have a specific point of origin, a specifically designated source. According to the EPA, nonpoint-source pollutants, on the other hand, are defined as any water pollution that does not meet the above definition. Nonpoint-source pollutants are, therefore, essentially everything else; they don't have a specific source. These types of pollutants are usually attributed to storm-water runoff. Point-source pollutants are the primary emphasis of the provisions of the CWA. As we will see, this shortsightedness has created a loophole in the CWA because the greatest amount of water pollution originates with nonpoint pollution sources, which have been neglected in this legislation.

The CWA is a massive regulatory system, and it has significantly expanded federal authority and oversight with respect to the environment. However, the CWA does take some measures to allow for minimal state autonomy. For example, states may administer the NPDES if they can demonstrate the ability and legal authority to execute such a system.[22] If the state fails to comply with the provisions of the CWA, however, such authorization may be withdrawn, ensuring the federal government's supremacy.[23]

In addition, "each state also is required to submit an annual report to EPA describing the quality of water within its borders and the progress made toward meeting the CWA objectives," regardless of whether it is under state or federal supervision.[24] The theory behind these provisions was fairly straightforward: States could administer the act, while the threat of cumbersome federal intervention would keep interstate competition from ruining the regulatory system by "racing to the bottom."[25]

Although allowing limited state autonomy in carrying out the federal mandate may seem like a move in the right direction, the CWA has many problems. The CWA has politicized water conservation choices and opened up far too much flexibility for political entrepreneurs. Under the common-law system we discuss below, courts were the regulating authority and possessed little flexibility, thus ensuring longstanding protection of individual property rights. One of the few flexibilities offered to judges under this system was to grant temporary "concession[s] to the public interest" from time to time, by simply delaying injunction deadlines for a limited period.[26] Concessions such as these, however, were not possible unless property rights were first secured.

Politicized choices carried out by politicians and bureaucrats have been enabled through the "legislated erosion of property rights."[27] Through federal conservation legislation, national regulations have gradually replaced common-law property rights, which has furthered this politicization. One result is that local stakeholders who better understand the necessities in preserving their perspective environments are removed from the decision process, which is turned over to federal interests.

When politicians who are far away from the problem are allowed to make decisions, mismanagement, corruption, and inefficiency increase. For example, in 1996, Atlanta had been "unauthorized" to pollute the Chattahoochee River but did so anyway. The Georgia State Environmental Protection Division, charged with overseeing the state's NPDES program, had the responsibility to intervene and end the pollution practices. Rather than working to correct the issue overall, the agency relied on the flexibility within the political system to be soft on the City of Atlanta, and it exacted only small fines rather than the maximum penalty for releasing the pollutants.[28]

The CWA has had some success. After all, an estimated 65 percent of U.S. waterways are now safe for swimming or fishing, compared to just 33 percent four decades ago.[29] "I've been at this game for a long time, and what I saw in wastewater discharges when I first started was so much worse than what I see today," says Ken Greenberg. As chief of the EPA's CWA compliance office for Region 9, covering the southern United States and 147 Native American tribes, Greenberg has experienced a fair amount of CWA history.[30] What's more, forty years ago, 30 percent of tap water samples exceeded federal chemical limits; today 90.7 percent of U.S. tap water meets all applicable health-based standards.[31]

In the traditional discussion of the CWA, these successes are held up as proof of the overall success of the act. These facts, however, do not paint a complete picture. According to the EPA's own assessment, at the current rate, it will take an estimated 500 years to remediate remaining impaired waters, even if no new impairments take place.[32] Furthermore, "additional waters have continued to be added to the impaired waters list at a significant rate."[33]

The CWA has been modified by a number of amendments since its passage. The 1977 amendments saw state assumption of the regulation of dredge or fill materials such as sand and gravel, alongside new effluent limitations and guidelines, state compliance deadlines, pretreatment standards, sewage treatment processes, and the redefining of many point-source pollutants.[34] The 1987 amendments attempted to address nonpoint-source pollutants by directing states to "develop and implement nonpoint management programs."[35] This simple direction, offered without consequence or incentive, has done almost nothing to prevent nonpoint source pollution. Since these 1987 amendments, although a few individual modifications occurred, no major program or requirement has been implemented.[36]

Additional problems with the CWA become apparent as it is implemented to help permit projects that might affect water. The CWA is easily manipulated in the right hands: Agencies use it to demand and maximize mitigation requests, and organizations use it to claim exaggerated impacts. The CWA is often redundant and incompatible with other environmental laws, such as the National Environmental Policy Act (NEPA). Finally, the CWA is routinely used to protect bureaucrats' backsides rather than to find better environmental results.

CWA and the Balance of Nature

The word *eliminate* is used thirteen times in the 1972 CWA in reference to pollution. In fact, section 1 of Title I of the CWA states as the "national goal that the discharge of pollutants into the navigable waters be eliminated by 1985." Eliminated? Anything not water but *in* water is a pollutant, so how exactly do we manage that? Why, in 1972, did we suddenly think we could eliminate all pollution from waterways within thirteen years? Calling for the elimination of pollution involves an inherent assumption that humans are not part of the natural environment and that no trace of them should be found in the natural environment, including in the water on which all life forms are dependent.

Under Section 404 of the CWA, impacts are taken into consideration to the "natural environment"—with humans excluded. As a result, social impacts and economic impacts such as those to families, homes, churches, and schools do not matter under the CWA. High costs resulting from compliance with the CWA are also irrelevant under the law; it's as if they do not even exist. The environment as defined by the CWA does not include the effects that a particular action may have on the *people* living in that area, just the "natural" things.

As noted previously, this balance of nature view is naïve and outdated. It is based not on science, but on emotion and erroneous thought. Political entrepreneurs, as they crafted the laws from this time period, took advantage of typical thought surrounding human roles in the environment. As a result, we are left with a CWA based on flawed science with objectives that are unattainable and primarily symbolic.

A Common Law Replacement

The growing environmental consciousness that generated the CWA neglected the history of environmental protections rooted in common law.[37] Heightened public fears, spread by well-publicized environmental catastrophes, led to overlooking or completely rejecting the already functioning, bottom-up approach to conservation that was being achieved (albeit less noticeably). This policy shift toward federal intervention represented a dismissal of centuries-old common-law evidence reiterating the capability of localized conservation.[38] In fact, the common-law system maximized environmental protections by securing property rights for the individual rather than society at large.[39]

Under common law, if landowners used hazardous chemicals that then seeped into the lands or water of their neighbors, their neighbors had the legal right to prove fault and receive financial compensation, an injunction, or both.[40] Under this system, environmental conservation is achieved because individual rights are guaranteed. What results is an incentive structure that prompts property owners to cooperate in minimizing harmful practices as a means to reduce the liability associated with inhibiting their neighbors' rights. Disrespecting their neighbors' rights would result in costly court-mandated changes that act as a deterrent against such behavior.[41] Thus, under the common-law system, property owners are incentivized to maintain healthful environmental standards.

Long before the CWA was passed, courts were upholding clean water standards. During the late nineteenth century, the Carmichael family of eastern Texas owned a forty-five acre farm just a few feet from the Arkansas border. When Texarkana, Arkansas, decided to build a sewage system that would discharge into a stream that ran through the Carmichaels' yard, the family sued the city. A federal court decided in favor of the Carmichaels. The court found that by dumping human waste onto the family's property, the city has created a "great nuisance" of a cesspool. Furthermore, it would prevent the family from the "use and benefit of said creek running through their land and premises in a pure and natural state as it was before the creation of said cesspool."[42] The court awarded damages and an injunction against the cesspool.

Perhaps the most important part of this case is the fact that the court could find no reason why the Carmichaels' property should not be protected against pollution. In fact, Judge Rogers noted, "I have failed to find a single well-considered case where the American courts have not granted relief under circumstances such as are alleged in this bill against the city."[43] Remember, this was the late nineteenth century, long before the passage of the CWA. Despite the fact that no top-down regulation of water quality existed, courts were enforcing clean-water standards based on private property rights. Peter Davis gives examples in which "Paper mills in Wisconsin routinely owned miles of downstream river property, knowing that otherwise they would be liable for violation of riparian rights."[44] Because businesses and cities knew they would be found liable for polluting their neighbors' water, they acted responsibly.[45]

In 1913 the right to use and enjoy clean water of one New York citizen was pinned against the rights of a local business. A pulp mill was contaminating a creek on which a farmer downstream relied for agricultural use. Under common law, even though the company had a right to use the creek, its actions were clearly illegal because they inhibited the farmer's use downstream. A legislation-based regulatory system might have allowed the pollution to continue at the expense of the farmer in effort to benefit society at large. After all, people depended on the pulp mill from both a consumer and producer perspective. At the time, however, common law governed, and the rights of the farmer were protected through both financial compensation and an injunction.

Although proponents claimed the CWA would make the waters of the United States cleaner, even "fishable and swimmable," it sometimes did just the opposite. For example, in 1972, the City of Milwaukee was polluting Lake Michigan with sewage. In response, Illinois sued the city for polluting the lake, a major source of drinking water for Chicago. When the case reached the Supreme Court, it was decided in favor of Illinois. The court held that, "a State with high water quality standards may well ask that its strict standards be honored and that it not be compelled to lower itself to the more degrading standards of a neighbor." According to the court, "While federal law governs, consideration of state standards may be relevant." Before the CWA was passed, states could voluntarily decide to have higher water quality standards than their neighbors, and those standards would often be enforced.

A few months after the court's decision, the CWA of 1972 was passed. In its wake, Milwaukee asked the court to reverse its decision. The Supreme Court found that "the establishment of such a self-consciously comprehensive program by Congress, which certainly did not exist when *Illinois v. Milwaukee* was decided, strongly suggests that there is no room for courts to attempt to improve on that program with federal common law." Thus, Illinois' higher standard of water cleanliness was wiped clean. The state would be forced to abide by the lower national standards set by the CWA. In the case of *Illinois v. Milwaukee,* the CWA resulted in making water dirtier rather than cleaner. This standard was upheld by another case in 1992 between Oklahoma and Arkansas. Oklahoma sued Arkansas for allowing pollution to flow across state lines. Although Oklahoma had higher standards than Arkansas, the court found in favor of Arkansas, deferring to the EPA's broad jurisdiction granted by Congress. These court cases demonstrate instances in which the CWA actually worked in direct opposition to its stated goal, lowering water standards rather than raising them.

The CWA was a significant shift from a successful common-law system to one that emphasized top-down federal regulation.[46] Rather than maintaining an emphasis on private rights, policies are left in the hands of distant politicians or bureaucrats, who are incapable of crafting policies broad enough to address all forms of environmental protection. This results in limited options, implementation of suboptimal policies, and conservation opportunities that fall through the cracks.

Total Maximum Daily Load

Some nonpoint-source pollution protections do exist, but they are very limited in scope. The original 1972 CWA contained a provision that required states to identify bodies of water that remained impaired by pollution even after point-source protection had been applied. For these waters, states were required to establish a TMDL of both point-source and nonpoint-source pollutants that the body of water could take in and still achieve water-quality standards.

This provision, Section 303(d) of the CWA, however, was not implemented until political entrepreneurs from environmental groups initiated a series of lawsuits and demanded action. As a result, the EPA and state entities have acted to fulfill Section 303(d), but 41,000 bodies of water still don't meet water quality standards.

The EPA is legally obliged to establish TMDLs, but establishing a TMDL doesn't ensure that a water body will be improved. The Congressional Research Service outlines the difficulties faced by the TMDL system:

> When a TMDL is developed, implementation is a major uncertainty. First, section 303(d) does not require implementation, and states' strategies for implementation vary widely. Only a few have laws requiring implementation plans, while many others rely on less structured policies. Second, a number of barriers to implementation can be identified. The most prominent is insufficient funding, but technical impediments such as insufficient scientific data also are a challenge. At the same time, factors that may aid effective implementation can be identified, including active involvement of stakeholders and governments, and adequate resources.[47]

There is no requirement for implementation, no funding, no incentive for state entities, and huge technical barriers. In addition, there is strong resistance from groups such as the agricultural industry, which essentially gets a bye, on many CWA requirements. It is not surprising that nonpoint-source pollutants have been described as the CWA's Waterloo.[48] With the minimal requirement of simply creating a plan of implementation, the CWA ignores the incentive structure surrounding the issue, giving local communities no incentive to actually implement their mandated plan; the pollution quite literally goes downstream.

The EPA's own timeframe projections in their National Evaluation of the CWA estimate that it will take a minimum of 500 years to cleanse the remaining impaired third of U.S. waterways, even if all other impairment ceases.[49] The report goes on to state, "the vast majority of our nation's impaired waters have no possibility of being restored unless the nonpoint sources are effectively remediated. Moreover, unless nonpoint sources are more effectively addressed, we will continue to see the number of impaired waters grow over time."[50]

TMDL in the Chesapeake Bay

The Chesapeake Bay TMDL gives insight into the lack of function of this system. The Chesapeake Bay has been under environmental scrutiny for the last twenty-five years, with program after program tried and discarded. A number of difficulties have contributed to the failure of past environmental protections. Some of the longest rivers in the eastern United States pass through several states on their way to dump into the Chesapeake Bay. As a result, reducing TMDL loads in the bay requires coordination among Delaware, Washington, D.C., Maryland, New York, Pennsylvania, Virginia, and West Virginia.[51] It is no surprise that the bay has high levels of nitrogen, phosphorus, and sediment. All these pollutants come from a variety of sources including high population densities, agricultural operations, urban and suburban storm-water runoff, wastewater facilities, and air pollution. One result is dead zones, where, according to the EPA, "fish and shellfish cannot survive," and necessary sunlight is inhibited.[52]

On May 12, 2009, President Barack Obama issued Executive Order 13508 directing the federal government to restore and protect the bay and its watershed. EPA officials interpreted the Executive Order as meaning that a TMDL was necessary. Described as the largest and most complex TMDL in history, the Chesapeake Bay TMDL required "extensive knowledge of the stream flow characteristics of the watershed, sources of pollution, distribution and acreage of the various land uses, appropriate best management practices, the transport and fate of pollutants, precipitation data, and many other factors."[53]

Local government environmental officials are arguing that the bay cleanup plan is too costly and simply undoable in the fourteen-year time frame the EPA has allotted.[54] They also point to substantial ongoing improvements.

The Department of Agriculture's Assessment of the Effects of Conservation Practices on Cultivated Cropland in the Chesapeake Bay Region" outlines the progress that local agriculture made in cutting pollution, long before any TMDL was imposed. A majority of cropland acres have structural or management practices to control erosion.[55] Almost half of cropland acres are protected by structural practices such as buffers or terraces.[56] With reduced tillage being used on 88 percent of acreage, the region has witnessed a 57 percent reduction in sediment pollution, a 36 percent reduction in nitrogen pollution, and a 39 percent reduction in phosphorus pollution.[57]

In comments on the EPA Draft Chesapeake Bay TMDL, the National Association of Conservation Districts pointed to these local successes, asserting that misguided federal regulation in the form of TMDLs will do far more harm than good and that encouragement of best management practices will be best reached through voluntary, locally led, and incentive-based efforts.[58] Along with the assertion that the proposed pollutant reductions are not economically or technically feasible, the association stated, "EPA is relying upon an untested and highly imperfect model of the Bay, including incomplete and incorrect information about agricultural practices in the region and their water quality performance."[59]

The costs of meeting the TMDL standards are immense. In Anne Arundel County, Maryland, costs may exceed $2 billion, said Ronald Bowen, director of the county Department of Public Works.[60] For a county with 540,000 people, those costs are immense.[61] Many counties in the region are facing costs of over $1 billion. Those are "big numbers and scary numbers," said Randy Bartlett, deputy public works director for Fairfax County in Virginia. Meanwhile, "we're cutting teachers, we're cutting police and we're cutting fire," he said.[62] Understandably, most people are not enthusiastic about cutting other services in favor of incremental water quality gains for people downstream.

George Hawkins, general manager at DC Water, warned that cleanup costs in urban areas will hit a point of diminishing returns. Where a decade ago it cost the Blue Plains Wastewater Treatment Plant about $100 million to reduce nitrogen discharges from 15 milligrams per liter (mg/l) of water to 5 mg/l, it will now cost $1 billion to reduce it from 5 mg/l down to 4 mg/l. With ten times the cost for one tenth of the reduction, costs may be substantially higher than environmental protection estimates originally predicted, and substantially less effective. The *Bay Journal* reports that the City of Frederick is already undertaking a $54 million wastewater treatment plant upgrade. That translates to a 100 percent cost increase for ratepayers. Storm-water upgrades will cost the city $600 million more.[63]

With multiple aspects of the TMDL aimed at the agricultural industry, it is not surprising that the American Farm Bureau Federation, Pennsylvania Farm Bureau, and other organizations filed suit to block implementation of the TMDL.⁶⁴ With the Pennsylvania federal district court's decision upholding the TMDL, the various organizations have expressed their intent to appeal the decision, to "go the distance."⁶⁵ Prior to the federal regulation, agriculture was making voluntary reductions and improvements; now, the rigidity of federal oversight has fostered resistance and obstinacy from farmers.

The Chesapeake Bay TMDL is a political entrepreneurship success story on many levels. The Obama administration was able to show it was getting something done. Agency personnel have a new, ongoing project to justify and increase their budgets. Environmental groups can point to the TMDL as a successful result of their lobbying efforts and of taking the federal government to court. But being a political success does not mean it will be an environmental success. With limited funding, possibly faulty assessments of cost, overwhelming technical complications, and unattainable benchmarks, the Chesapeake Bay TMDL is a case of political symbolism and likely environmental failure.

Regulatory Nightmares: Section 404

One of the most problematic provisions of the CWA is found in Section 404, which is enforced jointly by the EPA and the U.S. Army Corps of Engineers. Section 404 is meant to protect the nation's wetlands and waterways from degradation by dredged or fill material.⁶⁶ Like the rest of the act, however, Section 404 causes unintended consequences that often increase costs and delays for projects while accomplishing little or nothing. In fact, Section 404 regularly results in worse environmental outcomes than if it were not enforced at all. Some of the most harmful unintended consequences are detailed below.

Redundancy

In our experience, very few projects completely avoid wetlands and the need for a Section 404 Permit. As a result, nearly every major project faces the redundancy of conducting an alternatives analysis under both NEPA and Section 404. This redundancy results in increased costs and time delays as the differences between what the two laws require are sorted out. Often, taxpayers pay for a huge analysis under NEPA only to pay for a similar one during the

Section 404 permitting process. Speaking before Congress about permitting delays, Gretchen Randall, a public policy consultant, stated, "At a minimum, Congress should examine NEPA and the laws passed since then, such as [the Endangered Species Act], the Clean Air Act, and the Clean Water Act, and eliminate overlapping duplication in the legislation."[67] Redundancy has become so costly that agencies pre-allocate the same amount of funds to the Section 404 permitting process as they do to NEPA.

Alternative Incompatibility with NEPA

Section 404 requires an alternatives analysis of proposed projects that may impact wetlands or waterways. A previously completed NEPA analysis can be adopted by the Section 404 process, so long as the alternatives analysis is conducted exactly as Section 404 stipulates, and if the "chosen" alternative is the one deemed the "least environmentally damaging practicable alternative" (LEDPA). Problems arise when the alternative chosen under the NEPA process is not the LEDPA. NEPA is a procedural law with no teeth and does not force the choice of an environmentally friendly alternative. This is not so with the CWA. Under Section 404, the alternative chosen must be the most environmentally friendly.

Under NEPA, impacts to the human environment are considered. Under CWA, impacts to the natural environment are considered, but humans are excluded. As a result, social impacts and economic impacts are not considered in a Section 404 alternatives analysis. If, under NEPA, alternatives are dismissed from further consideration because they have high costs or impacts to homes or social centers (churches, schools, etc.), these alternatives must be reconsidered as practicable under Section 404. The environment as defined by the CWA does not include the effects that a particular action may have on the people living in that area.

On a recent transportation project we worked on, the Corps of Engineers required re-analysis of alternatives that were dismissed under NEPA because of high costs and social impacts. Although the preferred alternative under NEPA disrupted only fallow farm fields with a new road, the Corps of Engineers required a detailed analysis of an alternative that displaced over 200 families from their homes. In addition, although the preferred alternative under NEPA would cost about $50 million, the Army Corps required analysis of another alternative that would cost over $250 million and displace fifty families from their homes. Under Section 404, these alternatives were not considered unreasonable. More than a year and hundreds of thousands of dollars were spent reviewing these alternatives, which had been dismissed as ridiculous during the previous NEPA analysis.

Mitigation Not Considered for LEDPA

When determining the LEDPA under Section 404, emphasis is placed on avoidance and minimization, not on mitigation. Thus, a project alternative that could be mitigated to the point where little or no impact to resources would occur cannot even be considered under Section 404. The alternative must be the least disruptive *prior* to mitigation, even if the mitigation would reduce the impacts below what would be incurred with a different alternative.

Another project, with which the authors are familiar, will likely cause wetland impacts. Project proponents offered to put a bridge over all waterways and employ various mitigation strategies to reduce wetland impacts to less than one acre. Although this reduced impacts significantly, the Corps of Engineers required that the project first avoid and minimize impacts, putting the project in a suboptimal location and making the mitigation strategies proposed earlier unavailable. As a result, the project had more impact to the environment and failed to fully achieve the project's goals, resulting in a lose-lose for all involved. Again, the CWA fails to protect the environment through stringent regulation. Environmental benefits go unrealized because they technically cannot be considered under CWA's strict guidelines.

Arbitrary Decisions by Unqualified Staff

Section 404 stipulates that the Corps of Engineers, not the person performing the analysis, determine certain elements of the analysis. The determination of the "basic project purpose" is one such example. Despite little or no training in specific disciplines, Corps of Engineers employees must determine the basic purpose of the projects they oversee. Although this doesn't sound difficult, small differences in opinion can have huge impacts in time and expenditure.

For instance, in another transportation project we provided analysis for, the purpose of the project under NEPA was to build a larger road to facilitate "mobility and connectivity" in a section of a city that had developed without any larger roads to accommodate longer distance travel. An environmental impact statement was written for this project, but the Corps of Engineers, when reviewing the Section 404 permit application, decided that the basic purpose of the project for Section 404 purposes was to get from point A to point B. Transportation planners and roadway engineers tried to explain the basic concepts of mobility and connectivity in transportation planning to the Corps of Engineers representative, who was trained in soils science. He could not be swayed. In his mind, the purpose of a road is simply to get from one place to another. This arbitrary decision by someone unqualified to make the call subsequently required that additional alternatives be analyzed under Section 404 because the NEPA analysis was deemed insufficient. This determination added over a year to the project and cost hundreds of thousands of dollars.

Restoration Project "Impacts"

Ironically, changes to wetlands that would occur from wetland restoration projects are also considered impacts under Section 404. This causes serious problems for bureaucrats because the inherent assumption of the CWA is that projects destroy wetlands; they do not create or enhance them. How exactly do you avoid effects to wetlands and minimize those effects when the project is actually designed to enhance wetlands? How does the Corps of Engineers decide how much mitigation to require for the impacted wetlands when the whole purpose of the project is to affect and change wetlands?

These questions came to a head on a recent large-scale river restoration project, which would restore and enhance hundreds of acres of wetlands. The function and area of the wetlands would be improved, but the Corps of Engineers had no process in their bureaucratic arsenal to deal with improvements. As a result, they treated the project actions as impacts. Alternatives were considered that produced fewer restored wetlands, and impacted fewer low-functioning wetlands. The project personnel were required to estimate what the future functional benefits of the wetland restoration would produce. Mitigation was required to compensate for the benefits that the wetlands would receive. It was a completely useless and meaningless exercise in paperwork. The result was that the restoration project failed to restore as many wetlands as it could have without Section 404. Again, we see the CWA failing to protect (and enhance) the environment as it was intended to do.

Maximizing Mitigation

Section 404 requires the Corps of Engineers to consult with the public and agencies in determining whether to grant a Section 404 permit to the applicant. Sometimes, the public process and agency coordination that took place under NEPA are sufficient to meet this requirement. When political entrepreneurs have unresolved issues with the project, wish to delay or terminate the project, or decide that mitigation would be beneficial to their individual interests, the Section 404 process makes it much easier for these goals to be achieved.

The ease with which projects may be delayed by using the CWA is primarily due to the fact that political entrepreneurs have direct access to the decision maker—the Corps of Engineers representative. In fact, the representative actually invites interested parties to comment on ways the project may impact resources and how those impacts could be mitigated. Unlike the determination of the basic project purpose, the Corps of Engineers recognizes that it lacks institutional expertise on certain subjects and is deferential to agencies such as the U.S. Forest Service, state agencies, or environmental groups. As with the NEPA process, speculation, exaggeration, and requests for more information rule the day. These agencies and groups have only to convince the Corps of Engineers representative that they are right and the Corps of Engineers will write it in as a permit requirement. That is, the Corps of Engineers will stipulate that the applicant only gets a permit if the political entrepreneurs' requirements are met.

As a result, there is great power in the Section 404 process. Political entrepreneurs know that if they don't get concessions during the NEPA process, they can get them in the Section 404 process. Organizations and environmental groups lobby the Corps of Engineers representatives with tales of how terrible the project will be. Often, it is easier for Corps of Engineers representatives to give in to political lobbying and require additional concessions from the project. They work with other agency personnel on other projects and consider them allies in defending the environment, so when one makes a request for some mitigation, it is easy to say yes. After all, it is not the bureaucrat's money that is affected nor is his or her job in jeopardy. Why not require the deep-pocketed agency (e.g., Department of Defense or Department of Transportation) or an energy developer to pay a little more to help please your colleague?

During the Section 404 permitting process described earlier, the Fish and Wildlife Service demanded nearly 3,000 acres of mitigation lands be purchased and set aside for conservation by the project. The NEPA analysis noted only minor wildlife displacement effects, minor indirect effects, and minor cumulative effects. The request relied on an obscure conjecture from a scientist in Europe who noticed effects to birds within three kilometers of roadways in Wales. The political entrepreneurs at the Fish and Wildlife Service thought that surely the same impacts would happen in this case, despite vast geographic differences and species compositions, and so the project was responsible for this large-scale mitigation.

Originally laughed at by project personnel, the Fish and Wildlife Service pressed so hard for this so-called mitigation that the developer eventually spent millions of dollars to purchase over a hundred acres of land, enhance it with native vegetation, put it in a permanent conservation easement, and subject it to monitoring by the Corps of Engineers for a minimum of ten years. Meanwhile, project personnel are still wondering why the mitigation was implemented in the first place. Once again, the easiest way to get people to go away is to give them what they want. Because Section 404 requires only one person at the Corps of Engineers to be convinced of a project's need for mitigation, and because the Corps of Engineers or the requesting agency does not internalize the costs of requiring that mitigation, delays and costs can be maximized.

Problems with One Size Fits All Enforcement

When they face few political costs for the rules they impose, political entrepreneurs are able to impose costly rules. One example is the EPA's 2001 standards for arsenic levels in the United States. The new standard dropped the permissible levels of arsenic from 50 parts per million (ppm) to 10 ppm. This was a nationally enforced standard, with no options for local control over enforcement. Although it seems like a worthy goal, there was little research supporting the benefits. In addition, the costs to reach the EPA's goal were not uniformly distributed; if only a few people lived in a certain water district, there were fewer to share the cost of this new regulation than another water district where more people lived. Thus, these new arsenic standards were difficult for some districts to achieve.

In a larger district, the cost of such standards would be very low, anywhere from $1 to $25 annually per household, while in smaller districts it can cost $71 to $327 annually per household. Considering that 95 percent of the 54,000 community water systems in the United States were already in compliance with this standard, the EPA was levying high costs in a small number of districts. The most optimistic surveys suggested that it would save less than thirty statistical lives per year (one statistical life reduces the average number of deaths by one), but to many, that was an overestimation, and some experts even expected that no lives would be saved by this new standard. The argument was made that the money spent on arsenic standards could be better spent on things like mammograms and colon screening, which could potentially save more statistical lives.[68]

Local Control and the Supposed Race to the Bottom

In addition to knowing the individual cases better than a federal organization, local and state governments can react to citizens' concerns more quickly. In the famous case of the 1969 Cuyahoga River Fire cited at the beginning of this chapter, cleanup on the river had actually begun years earlier, and the quality of the water had improved significantly, partly due to the formation of the Cuyahoga Basin Water Control Committee.

Prior to federal intervention, local governments nationwide initiated similar conservation efforts. By 1966, state-sponsored environmental protection legislation had been passed across the nation, with every state enacting its own form of water pollution legislation.[69] A 2009 assessment from the Government

Accountability Office revealed that state enforcement programs had improved substantially since 1966.[70] As a Resources for the Future report noted, state and local efforts had been successful: "the results of the EPA's first National Water Quality Inventory, conducted in 1973, indicated there had been significant improvements in most major waterways over the preceding decade."[71]

The evidence above contradicts the theory behind the race-to-the-bottom argument that so many advocates of federal regulation use. Race-to-the-bottom is the theory that, if left to their own devices, states would decrease environmental regulation to appeal to industries in exchange for coveted tax dollars. By this reasoning, states would compete for the attention of large industries until there were no meaningful environmental standards at all. However, real-world experience does not support race-to-the-bottom predictions.

During the presidential administration of Ronald Reagan, some environmental regulation was put back into the hands of the states, and rather than a decrease in spending on environmental regulation, an increase was seen.[72] Even when adjacent states had *lower* environmental standards, states with higher standards did not drop theirs. In fact, the opposite of the expected result occurred. The states with *higher* environmental standards actually saw an increase in the spending on environmental efforts of adjacent states. In other words, when environmental regulation was put into the hands of the states, a race to the *top* rather than to the bottom had occurred.[73]

The political reason to prefer national rather than local regulation is not because of a supposed race to the bottom. It is simply much easier to lobby one level of government, in one location, than it is to fight for or against environmental regulations on state and local levels across the fifty states. Thus, most political entrepreneurs prefer federal over local control.

Inconsistent Enforcement

For any regulation to be effective, it must be enforced with some level of consistency. Inconsistency, however, creates large arenas for political entrepreneurs: those being regulated, those wishing to have their competition regulated, and those pushing for more regulation. When it comes to national water policy, the EPA's enforcement of the CWA has been inconsistent at best.

Multiple studies have found inconsistent enforcement to be one of the biggest problems with the CWA. In 2009 a Government Accountability Office report detailed key issues that impact enforcement of the CWA. This report found that "inspection coverage by EPA and state enforcement staff varied for facilities discharging pollutants within each region."[74] Across and even within regions, the "number and type of enforcement actions," "size of the penalties assessed," "criteria used in determining penalties," and "the regions' overall strategies in overseeing the states within their jurisdiction" varied widely. Reasons for such inconsistent enforcement included different "philosophical approaches" to compliance, "differences in state laws and enforcement authorities," "variations in resources available," "flexibility afforded by EPA policies," and "incomplete and inadequate enforcement data."[75]

The EPA intentionally allows for some leeway in how states and regional offices enforce the requirements of the CWA to "take into account local conditions and local concerns."[76] The Office of Enforcement and Compliance Assurance is EPA's headquarters and provides guidance for the agency's ten regional offices across the country. These regional offices then oversee implementation of CWA regulations either directly or by supervising state agencies that have been delegated to carry out enforcement. Despite this decentralized system, EPA has created guidelines for state programs, and, it "expects the regions to take a systematic approach to administering and overseeing the enforcement programs among delegated and non-delegated programs."[77] These expectations are not always met.

The EPA itself has noted that inconsistent enforcement is a problem. Lisa Jackson, former EPA administrator, testified to this effect before the House Transportation and Infrastructure Committee in 2009. She stated:

> Data available to EPA shows that, in many parts of the country, the level of significant non-compliance with permitting requirements is unacceptably high and the level of enforcement activity is unacceptably low. For example, one out of every four of the largest Clean Water Act dischargers had significant violations in 2008. . . . The government's enforcement response to these violations is uneven across the country. . . . [A] violation in one State results in the assessment of mandatory minimum penalties, while in another State, no enforcement action is

taken for the same violation. This situation creates a competitive disadvantage for States that are enforcing the law. We need to change this.[78]

As Jackson describes it, inconsistent enforcement of the CWA may be actually creating the race to the bottom that was feared before the CWA was enacted and that was its justification. She says the CWA is incentivizing companies to move to states where enforcement is more lax. This incentive structure punishes those states that consistently follow the law with negative economic outcomes while rewarding those states that violate provisions of the CWA. The clever entrepreneur, political or private, can use these incentives to simply move pollution across state lines or into different EPA regions.

Another EPA report, completed in 2011, found that even when states underperform, "EPA regions do not consistently intervene in states to correct deficiencies."[79] For example, from 2003 to 2009, North Dakota inspected 100 percent of its major CWA facilities, but failed to penalize any known CWA violators during this period. Region 8 stepped in and "increased their inspection coverage and file reviews," but compliance did not improve.[80] Louisiana, Alaska, and Illinois also ranked poorly on CWA enforcements during this time period and, like North Dakota, regional authorities were unable to correct the problem.

In other cases, the EPA takes a heavy-handed approach to enforcing the CWA. In September 2013 residents of Chicken, Alaska, were dumbfounded to see men carrying assault rifles and in full body armor with the word "POLICE" boldly displayed on their chests, speed by on ATVs. The men made their way to the edge of the Fortymile River and begin to take water samples.[81] Despite their gear, these men weren't paramilitary—they were EPA officials. The citizens of Chicken, a small mining town, said that they had always had rather friendly encounters with the EPA before that day, and they fittingly called this move "heavy handed." You might wonder why the EPA would take such an action with people who were cordial before; the simple answer is intimidation and budget justification. The EPA used its enforcement task force, normally reserved for hostile environments, to conduct routine compliance checks. The EPA claimed that this was done because of reports that the area was dangerous due to widespread drug use, but the troopers themselves claimed that they had no evidence the Chicken area was dangerous.[82]

Deep Pockets

In our experience, CWA requirements also vary widely, depending on how deep the pockets are of the project proponent. Those with more funding are often asked to mitigate to a higher standard than those with less funding. This is another form of inconsistent enforcement that makes it difficult for project developers to know what will be required of them under the law. On a recent transportation project, for instance, building a new road would impact about five acres of wetland. The Corps of Engineers and the Fish and Wildlife Service, knowing the "deep pockets" of the Department of Transportation, required the purchase of 100 acres of land to mitigate for this impact. This land, which had to be set aside as a conservation easement, ended up adding millions of dollars in cost to the project. Because the project had the ability to pay, strict mitigation requirements were imposed. Just a few miles away, another project was under way that would impact about eight acres of similar wetland in the same watershed. Because the project proponent did not have nearly as much funding, the mitigation requirements were not nearly as stringent.

Enforcement of the CWA is, if anything, consistently inconsistent. Projects across the country are regulated to different standards according to a variety of factors. As a result, the CWA has not been effective in doing its job. Unless reform occurs, incentives will continue to encourage businesses to move to states where enforcement is lax and polluting has relatively low costs.

Conclusions

Like other environmental laws of the time period, the CWA was based on the flawed balance of nature theory that excludes humans from the environment. The passage of the CWA marked a shift in the United States, from localized conservation efforts to federally mandated preservation. As part of the western liberal tradition, America's founders had embraced common law, which stressed the importance of individual property rights. Under this system, environmental preservation saw a variety of successes at the local level. A series of unfortunate environmental catastrophes in the decade prior to the CWA's passage resulted in a public oversight and/or rejection of these past achievements.

What resulted from this change in public perceptions was an increased regulatory system that lessened the importance of individual rights and increased political as opposed to private entrepreneurship. As such, incentives for both land and water conservation were altered, and individual landowners were no longer encouraged to protect their respective environments.

In the hands of political entrepreneurs, the CWA works like other environmental laws: It can be manipulated to achieve desired results that differ from the original intent of the law. It adds an additional and heavy layer of bureaucratic complexity and provides myriad opportunities to increase costs and delays for nearly any project. Given how regulators and other agency personnel often interpret the law, it may have worse environmental outcomes than if the law were not in effect at all.

8

The Endangered Species Act

IN THE EARLY twentieth century, Oregon Governor Oswald West and Francis Elliott, the first state forester of Oregon, resolved to form a state forest that would provide money for education in the state. Over several years, Elliott negotiated with the federal government to assemble Oregon-owned common school land within national forest boundaries into one contiguous block of state forest. In 1930 that land became the Elliot State Forest.[1]

Under the Oregon Constitution, timber revenues from the Elliot State Forest are invested in the Common School Fund to financially support public education. Management has focused on generating the greatest amount of revenue possible so as to maximize benefits for education.[2] Management for these purposes has been a success, with revenue checks going out to schools from timber harvest twice a year for over eighty years. In 2000 revenue checks to schools totaled $16.2 million. What's more, habitat has been continuously protected and managed.

In the late 1990s, political entrepreneurs discovered the Elliot State Forest. Because of almost a century of responsible timber harvest, the forest still contains some of the oldest of the old-growth forests in the Pacific Northwest. Valued by environmental spiritualists and wildlife alike, old growth has wide appeal to political entrepreneurs.

Fearing that the old-growth sections of the forest would be cut next, political entrepreneurs spent the next decade successfully using the Endangered Species Act (ESA) to reduce timber harvest on the Elliot State Forest. By 2013 timber harvest revenues could not cover the annual operating expenses of the forest, which posted its first deficit of $3 million.

As this book goes to press, the Elliot State Forest is not expected to fund any education in Oregon through at least 2016. Despite the fact that forest management over the last eighty-four years was able to balance endangered species protection *and* timber harvest, the ESA, in the right hands, was used to ensure the forest was managed only for species and not schoolchildren.

The Endangered Species Act

Of the tens of thousands of laws passed by the U.S. Congress, the ESA is one of the most widely known. For the average person, the ESA conjures visions of saving charismatic birds and animals like eagles, grizzly bears, and wolves. No other piece of environmental legislation embodies the balance of nature ideology as completely as does the ESA. It attempts to save all species, regardless of economic cost. It is based on emotion and symbolism, and it creates rich opportunities for political entrepreneurship.

Because listing species under the terms of the ESA may cause financial problems for those with endangered species on their property, the ESA may encourage "shoot, shovel, and shut up" mortalities to listed species. "Shoot, shovel, and shut up" is just what it sounds like. Private landowners have long been suspected of personally removing endangered species, then burying the evidence, when they discover sensitive species on their property. The motivation behind this is simple. If an endangered species is known to inhabit a parcel of land, restrictions on how that land is used are likely to follow. To avoid this, landowners have strong incentives to attempt to rid their property of the listed species. Numerous articles, both popular and scholarly, have discussed the incentives underlying the shoot, shovel, and shut up phenomenon, and this topic will be explored in more depth later in the chapter.

A second and closely related outcome of the ESA, which has compromised recovery for endangered species, is the deliberate, preemptive destruction of habitat. The motivation for this is similar. When a private landowner suspects an endangered species may inhabit an area, the landowner has an incentive to make the land less inviting to that species. There is also an incentive to extract resources as soon as possible in order to avoid becoming subject to the ESA. This idea is clearly illustrated by the case of the red-cockaded woodpecker, whose habitat on private lands in North Carolina is quickly clear-cut if there is any hint of that bird in the forest.

Some argue that the continued survival of the listed species (nearly 99 percent of those listed have survived) is evidence enough that the ESA is working. Staunch advocates of the ESA further argue that, on average, listed species haven't been protected long enough to expect recovery. In reality, over forty years, the ESA has achieved a recovery rate of less than 1 percent for listed species and by that measure is a resounding failure.

Species and the Balance of Nature

According to current interpretations of the ESA, we must save not only all endangered species, but all endangered subspecies, and even unique or endangered subpopulations as well. Under the ESA, species are not to be ranked as to their biological or ecological importance; instead, all species, subspecies, and populations are treated as equals, and all are to be saved, whatever the cost. The ESA is really an equal rights act for species, subspecies, and populations. This, however, is a political statement and not a biological reality.

Biologically, not all species, subspecies, and populations, were created, nor did they evolve, equally. Even major species can be lost without an ecosystem collapsing. There are, however, what are called keystone species, whose loss will completely alter or change an ecosystem—without such species there would still be an ecosystem, but it would be a different ecosystem.[3]

The beaver is an excellent example of a keystone species.[4] Beaver not only create and maintain riparian areas that are critical to hundreds of other species, but also alter the hydrology, energy flow, and nutrient cycling of aquatic systems. Beaver dams impound water and trap sediments that raise the water table, increase the wetted perimeter, and allow the extension of riparian communities into former upland sites. In addition, beaver dams regulate stream flow by storing water, reducing peak or flood flow, and augmenting low flows during summer. During dry periods, 30 to 60 percent of the water in a stream system can be held in beaver ponds.[5] By trapping silt behind their dams over thousands of years, beaver actually created many of the American West's fertile valleys.[6] Therefore, protecting beaver makes a lot more biological sense than protecting species like the grizzly bear or kangaroo rat, which are not critical to ecosystem control or function.

Moreover, some species, subspecies, and populations, are naturally rare, especially those on the edge of their range, the geographical areas throughout which they exist. The Canada lynx and North American wolverine are examples. The Canada lynx was listed as endangered in 2009, and in 2013 the U.S. Fish and Wildlife Service proposed listing the wolverine as threatened. These species are clearly rare today, but they have been rare in the lower forty-eight states for the past several thousand years. The Canada lynx and North American wolverine normally inhabit boreal forests, a habitat type that barely extends into the lower contiguous United States. Even if all people of European ancestry were removed and the western United States returned to its pre-Columbian condition, these species would be scarce. They are also predators, which are always less abundant than their herbivorous prey, which in turn are less abundant than the plants they consume. When species at one trophic level consume species at a lower trophic level, there is generally a 90 percent loss in energy. Thus, 100 units of plants can support only 10 units of herbivores, which, in turn, can support only 1 unit of carnivores—a trophic pyramid, with a large plant base and a small apex of carnivores. This explains why top predators, species that are not normally prey for other species and thus are at the top of the food chain, will always be rare, especially at the edge of their range. In addition, as the weather varies and the climate changes, the range of these species will contract and expand.

Trying to save top predators on the edge of their range, like lynx or wolverines, will always be a losing battle. Saving endemic species—species that for one reason or another have a restricted natural distribution—also makes little biological sense. Many of these species are rare because long-term climatic changes now favor other species, and only in isolated pockets do the former species remain. Biologically, saving these remnant populations makes little sense because large numbers of the same species are usually found in other areas where the habitat and climate are more favorable. For example, while Canada lynx and North American wolverine are rare in the northern Rockies, those species are common in Canada and Alaska.

The same, of course, is true of the grizzly bear and gray wolf; two species that have sparked intense endangered species debate in the western United

States. Although 50,000 to 60,000 wolves exist in Canada and Alaska, and thousands of grizzlies as well, U.S. agencies have spent tens of millions of dollars trying to save, and even reintroduce, remnant populations of wolves and grizzlies in the western United States. An endangered species act that was truly about species would cause managers to recognize that these are monies that could be better spent on species that are in danger of extinction globally.

The red squirrel provides an even better example of the misplaced priorities associated with using the ESA to save locally endemic populations. No one disputes the fact that red squirrels probably number in the millions, but in southern Arizona, red squirrels are rare because of environmental factors, not human ones.

In the distant past, when the climate was cooler and wetter, coniferous forests were common, even in southern Arizona. As the climate warmed and dried, however, coniferous forests retreated up-slope until those communities are now found only at the highest elevations in southern Arizona. Basically, these are mountain islands in a sea of desert. Because red squirrels live only in coniferous forests, and primarily in spruce fir forests, their populations have declined with their shrinking habitat. Today small, isolated populations of red squirrels live on Mount Graham and other secluded southern Arizona mountain ranges.

About 120 miles to the west of Mount Graham sits Kitt Peak, home to an array of immense astronomical telescopes. This is big science, annually supported by large tax subsidies and federal grants, but the scientists have a problem—Tucson. Astronomical telescopes are usually situated on remote peaks with dry climates to take advantage of the clearer air at higher elevations and the dark night skies. When it was first chosen as the site for an astronomical observatory, Kitt Peak was ideal, but over the years Tucson, fifty miles to the east, has grown dramatically. With that growth came light pollution. Tucson's street lights, signs, and other infrastructure now illuminate the night sky to such an extent that the efficiency of Kitt Peak's huge telescopes has significantly decreased and will continue to decline as Tucson continues to expand.

In response to Tucson's light pollution, the federal government and the universities it supports went looking for a site for another astronomical observatory, and they chose Mount Graham (sitting at 10,717 feet) in southeast Arizona. Although the area is remote, a paved road built years ago by the U.S.

Forest Service winds to the top of the Pinaleno Mountains. The area, including Mount Graham, is now crisscrossed with logging roads, summer homes, and campgrounds. In other words, this mountain is not pristine wilderness.

Building the new observatory on Mount Graham seemed like a sure bet. There were extensive roads and other development in the area, and the project had support from the federal government, big science, and the Arizona state legislature. Then someone realized that a remnant population of at most a few hundred red squirrels inhabits the area around Mount Graham. The ESA was invoked, and the fight began. Although two telescopes have been erected, squirrel proponents won a partial victory when the U.S. Court of Appeals blocked construction of a third. While our preference is to not mar the Pinalenos with additional development, spending millions of dollars to save a few hundred red squirrels makes little biological sense, especially when other species are in imminent danger with no funding allotted to help.

The standard argument for saving all the species, such as the subpopulation of red squirrels on Mount Graham, is that doing so protects biodiversity. In fact, saving biodiversity has become the centerpiece of many environmental campaigns. The important species, from an ecological rather than a political standpoint, however, need to be protected first because all species are not created equal. It is also more important to protect ecological processes than it is to save biological species per se.

History and Description

Attempts by the federal government to protect animal and plant species date back to the Lacey Act of 1900, which "allowed for the 'preservation, distribution, introduction, and restoration of game birds and other wild birds.'" Other early acts include the Migratory Bird Treaty Act of 1918 and the Migratory Bird Conservation Act of 1929; these acts were limited in scope and lacked the necessary teeth or breadth to have any significant effect on endangered species.

The federal government's first broad approach to protecting endangered and threatened species was to purchase sensitive lands and set them aside as habitat for the target species; this was enabled by the 1964 Land and Water Conservation Act, which created the Land and Water Conservation Fund. The fund was established "to assist in preserving, developing, and assuring accessibility to all citizens ... quality and quantity of outdoor recreation resources."[7] This act was, however, not directly targeted at preserving species but toward acquiring land that might incidentally be habitat for endangered species.

Congress's next major step in species conservation was the Endangered Species Preservation Act of 1966, the goal of which was "conserving, protecting, restoring and propagating selected species of native fish and wildlife."[8] Before this act was passed, the only method available to officials for carrying out that purpose was, again, to acquire habitat. The direct purpose of the 1966 act was preservation. The 1966 act also authorized the U.S. secretary of the interior to create the first endangered species list and allowed the Fish and Wildlife Service to spend up to $15 million per year in purchasing habitat. The 1966 act was limited in strength, however, because it prohibited only the "taking"[9] of listed animals within national wildlife refuge lands. The inability to protect species on private property was a key limitation of the 1966 act that was not addressed until 1973.

The Endangered Species Preservation Act of 1966 was followed by the Endangered Species Conservation Act of 1969. The 1969 act extended protection to endangered subspecies and prohibited interstate commerce in unlawfully taken endangered species. The 1969 act further required the secretary of the interior to list non-U.S. species threatened with worldwide extinction and prohibit their importation to the United States. In 1973 the Convention on International Trade in Endangered Species of Wild Fauna and Flora brought conservation to the center stage of international politics. The treaty produced by this convention, signed by eighty nations, monitored and restricted trade in certain designated plant and animal species. Ultimately, the weakness of previous species preservation acts, coupled with rising public and political focus on endangered species, led Congress to adopt a completely rewritten protection act.

The Endangered Species Act (ESA) of 1973, which passed in both the U.S. House of Representatives and the U.S. Senate almost unanimously, is presently the primary legislation governing the protection of endangered species in the United States.[10] It has been hailed by some as the "broadest and most powerful law in a century-old history of protecting fish, wildlife, and plants through national legislation."[11] Norm Dicks, former Representative for Washington's sixth congressional district, called the act, "the strongest and most effective tool we have to repair the environmental harm that is causing a species to decline."[12]

The ESA, a complete rewrite of the 1966 and 1969 acts, allowed the federal government to enforce prohibitions on the taking of endangered species on any lands of the United States, private or public.[13] In the ESA, Congress concretely stated that various species extinctions had been a result of economic growth. Congress spelled out the purpose of the ESA this way:

> To provide a means whereby the ecosystems upon which endangered species and threatened species depend may be conserved, to provide a program for the conservation of such endangered species and threatened species, and to take such steps as may be appropriate to achieve the purposes of the treaties and conventions set forth in subsection (a) of this section.

Furthermore, the ESA empowered administrators with funds for species protection, allowed flexibility in listing species as either threatened or endangered, and expanded species eligible for listing to include plants and invertebrates. The ESA also split enforcement authority and responsibility between the Fish and Wildlife Service and the National Oceanic and Atmospheric Administration (NOAA).

Under the ESA, the Fish and Wildlife Service and NOAA are directed to list endangered or threatened species and provide appropriate protection to listed species to help them recover. It limits both governmental and private actions on lands designated as endangered species habitat and removes all restrictions on the use of the Land and Water Conservation Fund for purchasing land. More important, the ESA made the taking of endangered species illegal everywhere in the United States, not just on federal lands.[14] With the passage of the newly rewritten ESA, private lands became regulated under endangered species legislation for the first time. This marked a significant change in the dynamics of conflict between private landowners (who typically want to use or develop their land for personal or economic reasons) and endangered species activists (who typically place protection of species ahead of private financial concerns).

The ESA provides protection for species that meet the definitions of endangered or threatened. An endangered species is defined as "any species which is in danger of extinction throughout all or a significant portion of its range."[15] A threatened species is defined as "any species which is likely to become an endangered species within the foreseeable future throughout all or a significant portion of its range."[16]

Once a species is listed, that species and its habitat, whether on private or public land, fall under the control of the ESA's provisions. Any activities that could adversely affect the listed species are prohibited, including killing, harming, selling, and transporting any members of the species. Subspecies and even specific populations of a species (for example, bald eagles in the contiguous United States) can be listed as endangered—even if the species is not endangered in other parts of its range.

Five specific factors are considered when the Fish and Wildlife Service assesses a potential species listing:

> 1) damage to, or destruction of, a species' habitat; 2) overutilization of the species for commercial recreational, scientific, or educational purposes; 3) disease or predation; 4) inadequacy of existing protection; and 5) other natural or manmade factors that affect the continued existence of the species.[17]

Listed species are protected under the ESA, which bans the taking of species. The specific definition of *take* is a fundamental source of the ESA's authority; it "means to harass, harm, pursue, hunt, shoot, wound, kill, trap, capture, or collect, or to attempt to engage in any such conduct."[18] The ESA's Section 7, titled "Interagency Cooperation," requires all federal agencies to comply with the law, targeting a major contributor to habitat modification and destruction: the federal government. Section 9 of the ESA is responsible for its jurisdiction over private land as well as public land, which was a novelty in environmental legislation at the time, and gives the ESA much of its power as "approximately half of listed species have at least 80% of their habitat on private land" and are, therefore, subject to the dictates of the ESA.[19]

The power of the ESA was further expanded with amendments in 1982, which made clear that decisions regarding the status of a species were to be "made solely on the basis of biological and trade information without consideration of possible economic or other effects."[20] Broad definitions of taking under the ESA and the unprecedented jurisdiction over any land where an endangered species can be found inevitably resulted in a similarly unprecedented number of legal battles.

Legal Cases and the Beginning of ESA Entrepreneurship

The first major legal case involving the ESA was *Tennessee Valley Authority v. Hill et al.* (*TVA v. Hill*) in 1978.[21] The Tennessee Valley Authority (TVA) is a federal utility corporation and the largest public power company in the United States. *TVA v. Hill* centered on the proposed Tellico Dam, a TVA project on the Little Tennessee River in Tennessee. There were at the time more than sixty other dams in the region. In its initial survey of the region concluded in 1937, the TVA had considered the Little Tennessee not worth damming. For years, the site was deemed to have the lowest potential of more than seventy sites investigated.[22] By 1950, all dams the TVA considered justifiable for "flood control, navigation, and power" had already been built.[23] The Tellico Dam project became the focal point of a fiercely contested legal battle by local citizens who viewed the project as economically unjustifiable and a misuse of TVA's power of eminent domain.[24]

A citizen coalition, the Association for the Preservation of the Little Tennessee River, formed in 1964, led the opposition.[25] This group stood little chance of halting construction until the passage of the National Environmental Policy Act of 1969,[26] the use of which led to an injunction halting construction until 1973, when the TVA produced the necessary documents to comply with the law.[27] That same year, the hitherto unknown snail darter, a small, snail-eating, bottom-dwelling fish, was discovered in the Tellico Dam Project area. The TVA's opponents recognized that under Section 7 of the ESA, if they could get the snail darter listed under the ESA, the TVA would be prohibited from taking any action that could jeopardize its critical habitat.[28]

In an act of clever political entrepreneurship, the dam's opponents decided to push the Department of the Interior to list the fish, not necessarily because of the fish's importance, but because it was the only tool they had left for fighting TVA and its powerful supporters in Congress. Zygmunt Plater, a former professor at the University of Tennessee College of Law who was deeply involved in fighting the dam, wrote the following illuminating passage in his 2013 book, *The Snail Darter and the Dam: How Pork Barrel Politics Endangered a Little Fish and Killed a River*:

> And how do we answer accusations of hypocrisy—the predictable claim that we don't really care about the fish, that we are just using it as a technicality. This is easy, in part: we have no choice other than the snail darter. If we are going to resist this destructive, uneconomic project, we'll use the only practical tool available—the same way they nabbed Al Capone on taxes, instead of on murder and racketeering. There's also the flip answer: "What good is a law if you can't use it?" Not to mention the dramatic facts the media will discover in our case might actually strengthen environmental policy by demonstrating the Endangered Species Act's public interest benefits.[29]

For Plater and his compatriots fighting the dam, stopping the Tellico Dam was required by the ESA; for those in favor of the dam, the snail darter became the symbol of misguided environmentalism used to halt economic progress.[30] The snail darter was listed under the ESA, which halted construction. The TVA, having already spent $100 million on the Tellico Dam project, took its case to the U.S. Supreme Court; ultimately, the Supreme Court held that the ESA prohibited the TVA from damming the river, regardless of the fact that construction had been under way before the ESA was enacted and before the snail darter was listed as an endangered species.[31]

The TVA dam opponents had little time to savor their victory as they were matched against one of the most effective political entrepreneurs to serve in Congress—Senate Majority Leader Howard Baker from Tennessee. Before the Supreme Court provided its decision, Baker had drafted a bill to create the Endangered Species Committee (known first as the "God Committee" and eventually the "God Squad"). Composed of seven Cabinet-level officials, it was empowered to review projects and determine whether a species could be destroyed due to pressing human needs.[32] In a third brief victory for the environmental group, and in spite of Senator Baker's intention, the Endangered Species Committee decided the Tellico Dam project "would be more economically and environmentally sound" without a reservoir.[33] Once again, the victory was short lived; TVA's ultimate victory was assured when Baker's coalition attached a rider to an appropriations bill in the U.S. House of Representatives making the Tellico Dam project immune from all protective laws and ordering the project completed.[34]

Three key precedents were set in this case. First, destruction of an endangered species' *habitat* is forbidden by the ESA. Second, the Endangered Species Committee can declare a species less important than human needs, as defined by Congress, in spite of the wording in the ESA placing species preservation before economic considerations. Third, political entrepreneurs can use the ESA to accomplish ends only tangential to saving species.

A second crucial case that tested the limits of ESA authority was *Babbitt, Secretary of Interior v. Sweet Home Chapter of Communities for a Greater Oregon* in 1995.[35] This case concerned the definitions and limits of taking under the ESA. In question in this case was interpretation of one term in the definition of *taking*[36]—what exactly did *harm* mean?

Bruce Babbitt, then secretary of the interior, issued a regulation interpreting the word *harm* to include any modification or degradation of the habitat of an endangered species.[37] This concept, explored to some degree in the Tellico Dam and snail darter case but not conclusively stated, was the key issue in *Babbitt v. Sweet Home*. A group of private landowners and citizens involved in the timber industry sued the U.S. Department of the Interior, asserting that the secretary of the interior had exceeded his authority and that his interpretation of the term *harm* to include habitat modification had harmed them economically. Specifically, the plaintiffs in the case alleged that the interpretation of harm in relation to the red-cockaded woodpecker and northern spotted owl had deprived them of the economic benefits of their private property.

The citizen group's arguments relied on several key points, some of which were procedural quibbles. The primary argument was that the precise language of the original version of the ESA included a definition of *take* that specified "destruction, modification, or curtailment of habitat or range" which was omitted in the final ESA.[38] The argument presented was that if Congress had intended to include habitat destruction or modification as forms of harm in the ESA, they would not have deliberately omitted that wording in the final version. Ultimately, the secretary of the interior's interpretation of the word *harm* was upheld using the "reasonable interpretation" rationale given precedent in *Chevron U.S.A. Inc. v. Natural Resources Defense Council, Inc.*.[39]

The decision of the court in this case reflects and clarifies the holding in *TVA v. Hill*: Habitat modification can be reasonably seen as harm to a species, despite the plaintiffs' argument that any minimal or unforeseeable harm therefore violates the law. The implication is that both direct and indirect acts that fall under the definition of *taking* in the ESA, and *harm* as defined by the secretary of the interior, are violations of that act. By extension, under this argument, any act (by anyone or anything) that harms the habitat of an endangered species can, therefore, be defined as a violation of the ESA.

If climate change/global warming[40] can reasonably be established to be causing loss of endangered species habitat, some have suggested the dictates of the ESA could be applied to any act driving climate change. It could even be argued that doing so is required by the interpretation of harm and supported by the Supreme Court's decisions in both *TVA v. Hill* and *Babbitt v. Sweet Home*.[41] Numerous human activities release carbon dioxide and other greenhouse gases, which has caused climate change, melting polar ice, and raising sea levels. When rising sea levels destroy or reduce habitat for endangered species, then the interpretation of harm as any modification or destruction of habitat could be construed to require intervention under the ESA. It can be argued that we are, by releasing carbon dioxide, violating the ESA on a daily basis.[42] This extension of the scope of responsibility of the ESA, to address emission of greenhouse gases and climate change, has not yet been used by the agencies overseeing the ESA as it is thought to be politically infeasible.[43]

Causes of the Endangered Species Act's Failure and Unanticipated Consequences

The Endangered Species Act fails to accomplish its designs and generates several unanticipated consequences. One cause of its failure is that it is based on the balance of nature ideology. Other causes are the incentives embedded in the act for mischief by private citizens and agency personnel.

Endangered Species and the Balance of Nature Ideology

Several failures can immediately be seen in any practical attempt to use the balance of nature ideology to address endangered species. The Earth itself has gone through innumerable cooling and warming cycles. Throughout those processes, the Earth has supported an amazing array of species in a wide range of population sizes. Which of these infinite balances should we be attempting to restore? Before the last Ice Age, or just after it? Before the arrival of Native Americans or before the arrival of Europeans? Or perhaps just before the Industrial Revolution? There has never been widespread consensus on which state or time period the authors of the ESA (or any other environmental act) had in mind. Without a reference or benchmark, any attempt to restore a species or the environment is ultimately doomed to a nebulous outcome. Applying the concept of a balanced natural state to environmental issues such as carbon emissions has at least a possibility of generating a real benchmark.[44] However, given the highly volatile and interrelated nature of species populations, the idea is particularly problematic when it comes to preserving species.

The Fish and Wildlife Service has, in its quest for a balanced natural state, even removed some nonendangered species to protect species on the endangered list in spite of a rather clear evolutionary advantage of the nonendangered species. Choosing which species will be allowed to succeed has raised questions about humans' appropriate role in managing nature. The counterargument is that humans are at fault for endangering the species in the first place (an argument with considerable merit in some cases) and we, therefore, are responsible for restoring it. If we accept the counterargument, to what state should species be restored? Is it appropriate to pick a winner in managing nature, at the expense of another, more evolutionarily fit species? For example, should we kill the barred owl to protect the spotted owl?

Pick a Winner

The northern spotted owl, listed as a threatened species under the ESA since 1990, is a highly territorial species that prefers old-growth forests with high tree canopies.[45] The Fish and Wildlife Service describes the bird as a "medium-sized, chocolate brown owl" that generally mates for life and has a life span of up to twenty years.[46] The decline in both range and population of this species to an estimated 2,360 pairs, primarily along the western coast of North America, is attributed to logging and a territorial war with the barred owl.[47] This conflict with the barred owl makes the case of the spotted owl and the debate it has generated of particular interest.

The barred owl, a large and aggressive owl with a broad diet, is primarily found in the eastern parts of North America. Since 1973 these owls have expanded into the northwestern portions of the spotted owl's habitat.[48] The conflict between these species has, much to the dismay of agency administrators, reduced the numbers of spotted owls. Barred owls have displaced and even bred with the smaller spotted owls.[49] Next to logging of old-growth forests, which is now restricted by the ESA, the Fish and Wildlife Service regards the barred owl as the chief and as yet uncontrolled threat to the continued survival of the spotted owl.[50]

This interspecies conflict (between species of owls rather than humans and owls) makes this case so unusual. In order to protect the spotted owl as required by the ESA, the agencies must, in effect, choose which species ought to be allowed to live in the overlapping habitat. The choice of which one is to be allowed to live is made for them by the ESA: The barred owl is not endangered and, therefore, must be the loser.

In order to fulfill its dictates under the ESA, the Fish and Wildlife Service has considered numerous plans, including both relocating some barred owl and killing others within the spotted owl's territory.[51] Of the choices, "the most effective method of 'removal' or 'control' appears to be by lethal means . . . shooting individual [barred] owls with shotguns."[52] While numerous small-scale studies have removed barred owls by various means, the Fish and Wildlife Service's decision to move forward with a plan for large-scale removal is unprecedented.[53] The agency has recently published an environmental impact statement detailing the motives behind its plan, which has been in the works since at least 2005, to remove or kill barred owls in areas of Oregon, Washington, and California.[54]

The Fish and Wildlife Service intends to remove, mostly by lethal means, roughly 3,600 barred owls throughout the test area using a "general approach involv[ing] attracting territorial barred owls with recorded calls and shooting birds that respond."[55] In a firm nod to environmentalism, however, the agency's specific recommendations do prescribe use of shotguns loaded only with nontoxic lead substitute in the process.[56] The plans have drawn fire from scientists and the public on ethical grounds, despite attempts by the Fish and Wildlife Service to make the plan more palatable and invitations to participate in the decision-making process with the help of an outside consultant ethicist.[57]

The decision made by the Fish and Wildlife Service to kill barred owls in order to save spotted owls is a controversial decision in which a government agency is choosing which species will live and which will die in an area. The spotted owl population, protected since 1990, has continued to decline despite significant decreases (up to 50 percent) in logging of old-growth forests in the spotted owl's range.[58] The incursion of the aggressive and more adaptable barred owl has been determined to be the primary cause of that decline; the Fish and Wildlife Service, acting under the dictates of the ESA but contrary to Darwinian notions, intends to remove the barred owl until the weaker spotted owls are safe. Despite occasional public and environmental outcry, this choice was fundamentally mandated by the ESA, which makes clear that preserving endangered species is the priority, regardless of economic or other considerations.

Preemptive Habitat Destruction and Constitutional "Taking"

Given the broad and nearly all-inclusive definition of *taking* in the ESA, once an endangered species takes up residence on private property, the owner of that property is effectively blocked from most development and extractive activities. The net effect is that private land has become public habitat for the species. The courts do not consider ESA restrictions on private property to be an exercise of eminent domain, which means that private landowners are not entitled to any compensation. The obvious incentive from expanding ESA provisions to private property is to extract the land's resources, develop it, or otherwise render that property inhospitable.

More formally, as University of Arizona economist Robert Innes put it, "the possibility of uncompensated takings gives landowners an incentive to develop their property early on in order to reduce the risk that it will later be appropriated for public use."[59] This may lead to early clear-cutting of forest habitat and other similar activities. As Robert J. Smith wrote, "the perverse incentive structure of the act accelerates destruction of the very habitat the act was designed to protect."[60] This unfortunate consequence has come to be referred to as "preemptive habitat destruction."

In their article, "Preemptive Habitat Destruction Under the Endangered Species Act," Dean Lueck and J. A. Michael[61] detail the impacts of this incentive system in North Carolina, where the endangered red-cockaded woodpeckers are most unwelcome guests. This species, listed as endangered since 1970, inhabits older living pine trees almost exclusively; by harvesting early, landowners can maintain value in their property while avoiding becoming subject to the ESA.[62] Lueck and Michael find that the ESA has led to significant destruction of habitat throughout North Carolina, as some landowners rush to harvest timber before the red-cockaded woodpecker can move in. Failing to do so and allowing even one of the woodpeckers to settle in "can cost up to $200,000 in foregone timber harvests."[63] According to Lueck and Michael's 2003 study, preemptive destruction of habitats in North Carolina alone resulted in the loss of enough habitat to support between twenty-five and seventy-six woodpecker colonies of two to nine birds each.[64]

Although Lueck and Michael point out that their findings in regard to the red-cockaded woodpecker do not necessarily hold for other species, the underlying incentive logically applies to any situation in which endangered species may take up residence on private land. Given that roughly 60 percent of the on-shore lands in the United States are in private ownership, the preemptive destruction incentive may have far-reaching impacts. Furthermore, roughly 95 percent of that land is considered rural open space and could be potential habitat for endangered species.[65] According to data published by the U.S. General Accounting Office, over 90 percent of listed species in the United States have habitats on non-federal lands (1994).[66] In a 1995 study, Lynn Dwyer, Dennis Murphy, and Paul Ehrlich found that 50 percent of federally protected species are found exclusively on private land.[67] Although the distribution across private and public lands of the roughly 4,000 species being considered for listing under the ESA is not known, a reasonable assumption is that they are distributed similar to species already listed.

Texas provides an illustration of the impact the ESA is having on private property owners. About 95 percent of Texas lands are privately owned. These lands support eighty-two federally listed species and 305 candidate species. The result is that many of the costs of protecting and conserving endangered species in Texas fall on rural landowners, who are primarily farmers and ranchers. In most cases, they must pay these costs without compensation or assistance.[68]

Endangered species policy makes it unlawful for private citizens to interfere in any way with an endangered species or its habitat, and it imposes severe penalties for those who do. Farmers violate the ESA if they plow their land and, in some cases, if they allow grazing on a pasture when an endangered species is present. In many cases, property owners are prohibited from cutting trees, clearing brush, using pesticides, planting crops, building homes, protecting livestock or themselves from predators, and building roads. They are often required to set aside numerous acres for no other purpose than to serve as habitat for endangered species.

The authors of the 1996 Environmental Defense Fund's report "Rebuilding the Ark: Toward a More Effective Endangered Species Act for Private Land" noted that the ESA discourages private landowners from protecting endangered species by creating, restoring, or enhancing habitat.[69] They explain:

> Their unwillingness often stems from the fear of new restrictions. They are afraid that if they take actions that attract new endangered species to their land or increase the populations of the endangered species that are already there, their "reward" for doing so will be more regulatory restriction on the use of their property. In its most extreme manifestation, this fear has prompted some landowners to destroy unoccupied habitats of endangered species before the animals could find it. One landowner, referring to the presence of red-cockaded woodpeckers on a small section of his property, announced, "I cannot afford to let those woodpeckers take over the rest of the property. I'm going to start massive clearcutting."[70]

The restrictions imposed by the ESA related to hosting an endangered species on private land are not considered a taking under the U.S. Constitution, however. A constitutional taking[71] refers to the phrase in the Fifth Amendment, which states, "nor shall private property be taken for public use, without just compensation." Under a strict interpretation of the Fifth Amendment, an owner whose property has been devalued because of government regulations could file a claim and be compensated; by most commonsense interpretations, the ESA (which effectively removes the potential economic gains of developing real estate or harvesting natural resources on a private property) is clearly a regulation that devalues property and therefore entitles the private landowner to compensation.

Opponents of the ESA's regulatory powers argue that the ESA is, in fact, a taking of private property without just compensation.[72] Proponents of the ESA's existing broad powers, however, argue that the ESA is a kind of zoning, and courts have not considered zoning laws to be taking subject to the Fifth Amendment.[73] Alternatively, enforcement of the ESA has, at times, been deemed a "fair exercise of the police power of the state."[74] This dispute remains unresolved.

In *Babbitt v. Sweet Home*, the court avoided the federal takings issue. The question before the court was whether the definition of *harm* to an endangered species included habitat modification. The court ruled that Congress meant to include habitat modification in the definition of harm and that the Fish and Wildlife Service's application of regulations prohibiting certain uses of private lands was consistent with Congress's intent. In its decision, the court avoided the issue of whether the requirements of maintaining the habitat constituted federal taking of the land, as that wasn't the central issue in the case.

Two cases have, however, suggested a possibility that, if the issue were pressed, the Supreme Court might be at least willing to hear a constitutional taking argument. The first case was brought by a South Carolina resident, Mr. Lucas, who bought two residential lots on a South Carolina barrier island with the intention of building single-family homes. After he purchased the property, however, the state enacted the Beachfront Management Act, which barred Lucas from erecting any permanent, habitable structures on his land. He sued the state agency, asserting the ban on construction deprived him of the "economically viable use" of his property and was, therefore, a taking. The court ruled in his favor and stated that government regulation of land that completely eliminates the economic use is per se a taking.[75] This case is not likely to benefit landowners whose property is devalued by ESA regulations. It required that all of the property's economic value be taken, not just a part, as is the case with ESA regulations.

The second case, *Dolan v. City of Tigard*, was filed after the city of Tigard, Oregon, denied Florence Dolan a building permit for her property unless she dedicated 7,000 square feet of land for storm-water management and a park. The Supreme Court ruled in her favor and said the city was attempting to force Mrs. Dolan "to bear public burdens which, in all fairness and justice, should be borne by the public as a whole." The majority opinion further stated that the city should not try to avoid "the constitutional way" of paying for what it wants and that the Fifth Amendment should not be relegated to the status of "poor relation" compared with other amendments in the Bill of Rights.[76] Despite these rulings, which seem to indicate some possibility of defining application of the ESA as a federal taking, our research has not revealed any cases specifically pursuing that issue.

Because courts do not recognize ESA regulations as a federal taking, no judicial remedy is available to private landowners for economic losses if endangered species inhabit their lands. Because of this, the ESA creates an incentive system in which landowners are rationally inclined to preemptively destroy habitat in order to avoid economic losses. They are, otherwise, effectively deprived of their property by the ESA. As the National Association of Home Builders Developer's Guide to Endangered Species Regulation put it:

> The highest level of assurance that a property owner will not face an ESA issue is to maintain the property in a condition such that protected species cannot occupy the property ... This is referred to as the "scorched earth" technique.[77]

Given the large percentage of on-shore land in the United States in private hands, and over 90 percent of listed species having some habitat on those lands, it is inevitable that some conflict would arise between private landowners' rational self-interest and the ESA. It is also little wonder that, despite the general good intentions of people, the ESA has struggled and occasionally failed altogether to preserve species. "Most landowners want to have, and help, rare species on their land ... but the government's harsh penalties and the dire financial consequences that can come with finding an endangered species on your property are encouraging landowners to make their land inhospitable to endangered species."[78]

Shoot, Shovel, and Shut Up

Endangered species are also at risk from a more direct threat than the destruction of habitat when some private landowners simply shoot, shovel, and shut up. This phrase, most commonly used in reference to the conflicts between wolves and livestock in the western United States, may seem ruthless and abhorrent. Although it is illegal, shoot, shovel, and shut up makes sound financial common sense to landowners threatened with encroaching endangered species. The slogan is so popular among certain groups of people that one can purchase t-shirts and bumper stickers sporting the slogan.[79]

The idea is simple and, it is similar to the preemptive habitat destruction incentive. If an endangered species appears on private property, the landowner can either suffer the financial impact of the ESA or the landowner can kill the creature, get rid of the evidence, and never tell a soul. This is another case of the inadvertent incentive structure created by the ESA leading to truly unfortunate, yet not unpredictable, consequences. This is particularly true among ranchers in Idaho, where the gray wolf has been reintroduced. The conflict between private citizens and listed species has been an illustration of the underlying failure of the ESA to align private citizen motives with species preservation. The Fish and Wildlife Service has made an effort to curtail shoot, shovel, and shut up incentives by offering compensation to ranchers for livestock losses by wolf predation. Some sources suggest that getting compensation is more effort than simply killing the wolves.[80]

It is unfortunate that the incentive structure created by the ESA discourages preservation of species by private citizens. Quietly killing and burying endangered animals is easier and prevents losing control of private property. Those who are conservation minded, love the outdoors, and would like species to flourish—even these individuals admit that the threat of being subject to the ESA leaves practical landowners with little choice but to take the more direct if illegal approach.[81]

Opportunities for Political Entrepreneurship

As we noted earlier in the discussion about the Tellico Dam, the ESA provided a new arena for political entrepreneurship. In the case of the dam, the ESA provided a way for citizen groups to fight an entrenched public works bureaucracy and a powerful congressional leader. That battle was just the first in an ongoing conflict. Today, the ESA is a tool for political entrepreneurs to use to fight oil and gas development, grazing, utility-scale solar and wind farms, off-road vehicle use, coal mining, mining in general, timber harvest, hunting, outfitters, guides, or anyone else using public land and often private land. It is, in fact, a land-use tool cloaked as a species preservation tool.

In the case of the gray wolf, critics question whether the wolves should have been listed in the first place because the Canadian and Alaskan gray wolf populations (estimated at 56,000 to 58,000 wolves) are not in danger of extinction.[82] The successes of the Canadian and Alaskan populations are, however, not a part of the consideration for the listing of, for example, the Michigan or Minnesota population. Each population is considered separately.[83] This conflict has led to claims that the interest groups who advocate listing species have ulterior motives.[84]

To address the political opportunism that many perceive in how the ESA is used, members of Congress often introduce bills to restrict the reach of the ESA. During a debate over one such proposal, "Representative Mike Simpson (R-ID), chair of the House interior and environment appropriations subcommittee, a main proponent of the defeated measure, said he intended to send the feds a message that the Endangered Species Act had gone off course and needed to be righted."[85] Rep. Simpson went so far as to say, "The Endangered Species Act has become not so much about saving species as it has been about controlling land and water." To illustrate his case, Rep. Simpson used the Fish and Wildlife Service's 2009 listing of slickspot peppergrass, an Idaho native grass species, due to moderate decline.[86] Simpson accused the Fish and Wildlife Service of listing the grass "to prevent cattle grazing on public lands . . . That's the only reason that the slickspot peppergrass is really listed."[87] Former Pennsylvania Senator and Representative Rick Santorum has characterized the ESA "as a 'radical ideology' that puts 'critters above people'"[88]

Although these allegations are seemingly harsh and perhaps overreaching, similar accusations have been made in regard to the massive curtailment of logging required to protect both the spotted owl and the red-cockaded woodpecker.[89] Other proposed species listings have aroused the ire of petroleum companies and some media personalities, who allege that activists use the ESA as a means to prevent development, deliberately hurt industry, or generate revenue through litigation.[90] Some claim that lawyers (rather than scientists) and policy (rather than scientific evidence) decide what species are placed on the endangered species list and to what degree they are protected.[91] The idea that the dictates of the ESA imply a requirement to control emissions and address climate change, and the fear that environmental groups will use the ESA as a blunt instrument to attack industries, are more concerning to opponents. The suggestion that the ESA carries a requirement to address climate change presents a fertile ground for litigation.[92] Said Doc Hastings, chair of the House Natural Resources Committee, "one of the greatest obstacles to the success of the ESA is the way in which it has become a tool for excessive litigation. Instead of focusing on recovering endangered species, there are groups that use the ESA as a way to bring lawsuits against the government and block job-creating projects."[93] He continues: "Due to rigid timelines, vague definitions in the Act and the propensity of some groups to sue the agencies as a way of generating taxpayer-funded revenue, the ESA has become taken over by lawsuits, settlements and judicial action."[94]

In response to criticism of the ESA and activist groups, supporters point out that in the ESA, Congress has explicitly placed protection of endangered species before economic concerns; the impact on citizens or companies is not the concern of the ESA. They further argue that those who advocate for species do so out of a concern for the public interest, not for private interests.[95] In other retorts to claims of ulterior motives, such as prevention of logging, the answer is, again, simple: Those areas are critical habitat and, as such, protection of them is protection of the species. Some activist groups excitedly argue application of the ESA to climate change is required due to impacts on polar bears and other coastal creatures.[96]

The accusation that environmental political entrepreneurs use the ESA, among other legislation, as a tool to pursue hidden agendas of general preservation and curtailment of industry is unresolvable; no amount of argument from either side will conclusively answer if there is a hidden agenda. But, just as we expect private entrepreneurs to use whatever tools they have available, we expect environmental entrepreneurs to use the political tools they have available.

Ineffectual

The final subject addressed here, and one of the most hotly debated, is the effectiveness of the ESA. As David Ridenour, writer for the National Center for Policy Analysis, put it in 2005:

> In the 32 years the ESA has been on the books, just 34 of the nearly 1,300 U.S. species given special protection have made their way off the "endangered" or "threatened" lists. Of this number, nine species are now extinct, 14 appear to have been improperly listed in the first place, and just nine ... have recovered sufficiently to be de-listed. Two species—a plant with white to pale-blue flowers called the Hoover's Woolly-Star and the yellow perennial, Eggert's Sunflower—appear to have made their way off the threatened list in part through "recovery" and in part because they were not as threatened as originally believed. A less than 1% recovery rate isn't good.[97]

More recently, a statement by Doc Hasting, chair of the House Natural Resources Committee for the House, echoed Ridenour's statement: "the purpose of the ESA is to recover endangered species—yet this is where the current law is failing—and failing badly. Of the species listed under the ESA in the past 38 years, only 20 have been declared recovered. That's a 1% recovery rate."[98] (The counterargument is predictable: 99 percent of the species given protection have either recovered or are still on the lists; because they haven't gone extinct, the ESA is therefore very effective.)[99]

The response is that defining success as continued existence of a species is a poor standard and, furthermore, that any recovery or continual viability of listed species is not necessarily attributable to the ESA.[100] The recovery of numerous birds, including bald eagles, can be attributed to banning DDT, which had nothing directly to do with the ESA.[101] Ridenour points to several other recoveries unrelated to the ESA, such as the peregrine falcon, Aleutian goose, American alligator, and gray whale.[102] According to Ridenour in 2005, only "36% of the species on the endangered and threatened lists are currently believed to be stable or improving—meaning that 64% are declining."[103]

There is considerable disagreement regarding the success rate of the ESA, and even how to define success. Wildly contradictory statistics illustrating the low rate of delisting and acknowledgments by the Fish and Wildlife Service can't dissuade activist groups from proclaiming a heartwarming success of the ESA.[104]

Another argument presented by ESA advocates regarding the apparent low success rate for species being delisted is that "the great majority of species have not been listed long enough to warrant an expectation of recovery."[105] This argument asserts that the period necessary for recovery has not been reached, and attempting to measure success before that time is not informative. Of the 1,396 species listed as of 2011, the average time on the list was twenty-one years of the forty-two years for recovery.[106] These advocates point out that, given that the average period elapsed is only half expired, expecting any degree of success is unfair; the same paper, however, proclaims an overwhelming success rate.

In a 2012 study of 110 species chosen by the Center for Biological Diversity, the authors claim that "90% of species are recovering at the rate specified by their federal recovery plan."[107] That sweeping statement implies that the 90 percent success applies to the ESA in general, rather than to just the 110 species chosen by the authors out of nearly 1,400 listed species.[108] A sample of less than 10 percent of a population of endangered species, coupled with vague and misleading wording, raises doubts about the validity of their claims. Others point out that over 80 percent of species have recovery plans yet were not examined in the center's study.[109] The study and other attempts to quantify the success of the ESA are often perceived as contradictory, misleading, or biased, and all too frequently lacking in any scientific basis. Due to controversy regarding results, the actual effectiveness of the ESA is impossible to quantify.

An overall low success rate of the ESA would be unsurprising, given the enormously complicated and interrelated nature of ecosystems, the unfortunate incentive system created by the ESA, and limits to agencies' knowledge, ability, and authority. One cannot expect the Fish and Wildlife Service or NOAA to be omniscient; furthermore, given the increasing rate of decline in many species leading to new listings, the pressures placed on the agencies in carrying out the dictates of the ESA are only growing. Providing still more difficulty for the agencies, "litigation has become one of the greatest obstacles to the success of the ESA. Instead of focusing on recovering endangered species, groups are using the ESA to file hundreds of lawsuits."[110] It is little wonder, given the complexity of the task and inertial force of species decline and habitat pressure, that the ESA has had either little, or unclear, success.

Conclusions

Troubled ecosystems are exceedingly complicated, and solving the problems they present is difficult. These problems are impossible to solve, however, without a realistic ecological view of how nature works. The arguments against the ESA and questionable results it has yielded illustrate the ongoing difficulty and lack of consensus regarding demonstrated success. It is unrealistic to expect a complex political environment consisting of disparate, rationally self-interested individuals with divergent goals to agree on how to protect a complex ecosystem composed of fragile and nebulous populations of endangered species, or even how to measure their success.

With so much of the agency's resources devoted to dealing with litigation, it is little wonder that by 2009 federal and state expenditures related to the ESA had approached $1.5 billion, and yet still no consensus regarding demonstrable success has been reached.[111]

The facts of species decline and the difficulty faced by the agencies who carry out the intent of the ESA are apparent: With less than 1 percent of listed species having recovered enough for de-listing, the ESA is struggling at its chief purpose.[112] While the ESA is not a complete failure—the vast majority of listed species are at least hanging on to existence—there is a great deal of room for improvement. It cannot be fairly said that the agencies tasked with enforcing the ESA (i.e., Fish and Wildlife Service, NOAA) are at fault for this lack of success. These agencies are working against unintended incentives, the weight of increasing species decline, the addition of new species to the protection list, budgetary restrictions, and knowledge limitations. It is remarkable how well the agencies have done, given all that is working against them.

To achieve a more efficient and effective approach for the preservation of species and biodiversity, several issues must be addressed in reforms to the ESA. Reforms are needed to properly align the interests of private landowners with preservation of endangered species. Despite the ill-informed opinions of some who think landowners need not be compensated to set aside habitat for listed species, that could well be the most cost-effective approach—direct federal or state payments to landowners for long-term hosting of endangered species on healthy habitat. Fundamentally, if the goal of the ESA is to save endangered species from extinction, any cost-effective approach deserves consideration.

Excessive litigation detracts from the success of the ESA and may be used for ulterior motives including preventing development. Critics of litigant groups and the conduct of the ESA assert that "the implementation of the ESA too often goes beyond the original intent of species recovery and is instead used to block and delay job-creating economic projects and activities."[113] Whether groups use the ESA for ulterior motives or not, litigation drains resources from the Fish and Wildlife Service and NOAA, resources that could be more effectively used to preserve species. When "millions of dollars are wasted on frivolous lawsuits, resources are diverted away from true species recovery."[114]

Regardless of controversies regarding efficiency, effectiveness, and misuse of the ESA in the pursuit of ulterior motives, or the adverse incentive structures inherent in the ESA, many people support the preservation of endangered species. The question is not whether this worthy goal should be pursued. What is in question is how we have been pursuing this goal for the last forty years. As numerous litigants and activists have pointed out, the ESA has had some positive effects in preserving species. Exaggerations by these groups, however, mask the true situation and prevent meaningful progress toward a more effective and efficient ESA. Acknowledging the successes alongside the failures and addressing the underlying incentives for rationally self-interested people is required to steer the ESA toward effective methods of preserving species, rather than leaving it as a highly political environmental tool that is all too frequently abused.

9

The Wilderness Act

THE MUIR WOODS in California are named for the now revered conservationist who called for the preservation of his "cathedrals" in the mountains and led a comprehensive campaign to preserve what is now Yosemite National Park. Although John Muir is primarily remembered as an explorer and conservationist concerned with environmental degradation, deforestation, and disruption of wild places, the actual story is less about conservation and far more about the actions of an entrepreneur scanning the policy horizon and seeing an opportunity to capture his preferred outcome from those with the power to enforce it.

Muir devoted his life to creating parks and wild areas that did not contain people, with the exception of naturalists and tourists. Many of his ideas took shape when Muir visited and lived in California's Yosemite Valley. After noticing "the Miwok Indians growing crops, white settlers raising sheep, and miners seeking gold and other minerals," Muir decided "the other occupants had to go."[1] Although Muir opposed Indian oppression in his home state of California, he fully supported the extraction of Miwok Indians from Yosemite and even referred to them as "'dirty,' 'deadly,' and lazy."[2] For Muir, it was more important to maintain the balance of nature than to allow the Miwok Indians to live off the land. Muir's ideology about the balance of nature within national parks was so influential that "the Yosemite model spread to other national parks, including Yellowstone, where the forced evictions killed 300 Shoshone in one day."[3]

While touring Yosemite with Teddy Roosevelt, Muir asked the president to recognize that some areas needed conservation. Muir convinced him that without formal protection, the awe-inspiring landscapes of Yosemite were in danger of degradation and destruction. Muir's interaction with the president suggests that preserving Yosemite was a necessary and critical step in achieving his vision of a permanent wilderness in the Yosemite Valley at the expense of others.

What the Law Intended

Howard Zahniser, another preservationist and political entrepreneur, wrote the text of the Wilderness Act. In the forward to a biography of Zahniser, William Cronon wrote, "Howard Zahniser arguably made a greater practical contribution to the protection of wilderness in the United States than any other single individual in the past hundred years."[4] Zahniser was the executive secretary of the Wilderness Society from 1945 to 1964.[5] During that time, he worked in Washington, D.C., to promote and lobby for the protection of wilderness areas. In 1956 he wrote the first draft of the Wilderness Act, and for the next eight years, he lobbied for its passage. He even famously had a coat with extra pockets filled with copies of his bill and other handouts for everyone he met on Capitol Hill.[6] Zahniser never saw his efforts pay off; he died a few months before the law's passage.[7]

Congress passed the Wilderness Act in 1964, after eight years of debate, sixty-six rewrites, and eighteen public hearings. The act created a legal definition for wilderness and set aside nearly 9 million acres of federally owned land throughout the United States as a part of the National Wilderness Preservation System. The wilderness areas were described by the law as areas "where the earth and its community of life are untrammeled by man, where man himself is a visitor who does not remain."[8]

The purpose of the act was "to assure that an increasing population, accompanied by expanding settlement and growing mechanization, does not occupy and modify all areas within the United States and its possessions, leaving no lands designated for preservation and protection in their natural condition."[9] President Lyndon B. Johnson invoked maintaining the balance of nature when, upon signing the bill, he remarked, "If future generations are to remember us with gratitude rather than contempt, we must leave them something more than the miracles of technology. We must leave them a glimpse of the world as it was in the beginning, not just after we got through with it."[10]

The goal of the Wilderness Act was to protect federal wilderness in national parks, forests, wildlife refuges, and other public lands from the influence of man. As a result, wilderness areas became land that either restricted or completely prohibited most human activity. New development is forbidden in wilderness areas, as are most forms of recreation. There are no roads, no cars, and no buildings. Except in cases where previously valid legal claims exist, mining and timber harvesting are not allowed. Grazing is permitted only in areas where it was permitted before the wilderness area was designated. Even when activities like grazing, mining, and timber are allowed within a given wilderness area, they are subject to the regulation of various administrative agencies. The Wilderness Act was framed under the assumption that eliminating human actions in wilderness areas would help maintain an area's natural quality—humans and the environment were seen as mutually exclusive concepts.

Balance of Nature

The goals of The Wilderness Act of 1964 are based on the balance of nature ideology. As we noted earlier, this non-scientific idea is popular with environmental groups, the media, and policymakers, and romantic notions of a pure, pristine nature untouched by human hands still exist. Wilderness, as defined by the Wilderness Act, is little more than an idealist's fabrication based on cultural myths. William Cronon describes a common conceptualization of wilderness in the United States "For many Americans wilderness stands as the last remaining place where civilization, that all too human disease, has not fully infected the earth. It is an island in the polluted sea of urban-industrial modernity, the one place we can turn for escape from our own too-muchness." The balance of nature ideology applied to wilderness has resulted in wilderness restrictions that are not ecologically justified. Landres et al. highlights some examples:

> Lightning-ignited fires typically are allowed to burn, but human-ignited fires are not, even if their ecological benefits to the health of wilderness ecosystems would be identical. Or bare ground may be mitigated if attributed to humans or domestic livestock but not wild ungulates.

The management model required by the Wilderness Act keeps associated resource management shackled to an unsound balance of nature ideal. In other words, while land managers often understand that the balance of nature is an undefined myth, they are not allowed to change their management practices with the advance of ecological science. Timothy Profeta, at the 1996 Duke Environmental Law and Policy Forum, attacked these assumptions within the Wilderness Act, writing, "Unfortunately many of the laws designed to regulate ecological resources were passed when the 'Balance of Nature' paradigm was king and have not been redrafted to comport with advances in ecology."

The Wilderness Act is a "law predicated upon a fantasy" and bolstered by the balance of nature paradigm. Despite the Wilderness Act's purported intentions of protecting the environment, the reality is that the law fosters unnatural conditions by "[divorcing] indigenous people from their environments and, instead, [preserving] it as a place of recreation for those who live elsewhere."[11] The balance of nature assumption within the Wilderness Act continues to veil interests of wilderness management, preservationists, and groups benefiting from positive externalities associated with wilderness designations. Unfortunately, these incentives sometimes come at the economic expense of rural communities and may sacrifice the recreational, scenic, historical, and ecological goals designated wilderness areas are intended to protect.

As we examine the history of political entrepreneurs in relation to wilderness, keep in mind that there was no real public demand for wilderness. It was essentially a solution in search of a problem. With its unintended consequences and management headaches, the Wilderness Act is a vivid reminder of what can happen when political entrepreneurs desire private benefits from public costs.

Competition Between Agencies

While the burgeoning environmental movement contributed greatly to the creation of the Wilderness Act, there's more to the story. The political entrepreneurship of federal agencies had a significant role in the creation of wilderness areas, specifically a competition between two federal agencies: the U.S. Forest Service and the U.S. National Park Service.

Until the end of the nineteenth century, the General Land Office, under the U.S. Department of the Interior, managed most of the nation's public lands.[12] Its primary mission was selling land to the public. When the federal government first started to create the forest reserves, which became the national forests, the office was not the appropriate agency to manage them. Seeing this problem in 1905, the U.S. Congress transferred millions of acres of public lands to the Forest Service under the U.S. Department of Agriculture.

As a result of the transfer, the Forest Service budget increased from $439,000 in 1905 to nearly $3.6 million in 1908. Managing public lands, it turned out, was a great way to get funding from Congress.[13] The growth of the budget motivated the Department of the Interior to attempt to reclaim some of the General Land Office lands it had lost. An opportunity to do this arose in 1916, when Congress created the National Park Service[14] and placed it under the authority of the Department of the Interior, which used the National Park Service to retake lands from the Forest Service and the Department of Agriculture. The Department of the Interior did this by lobbying Congress to transfer lands, and from 1920 to 1928, nearly 600,000 acres of land were transferred from the Forest Service to the National Park Service.[15] With more land came more funding from Congress and greater job security.

During this period of reallocation, Forest Service officials realized that they could prevent their land from being incorporated into the parks if they designated it as "wilderness area." In other words, the Forest Service began designating wilderness areas as a way to neutralize the threat of the National Park Service taking them; pressure from the Park Service led the Forest Service to begin designating certain areas to be administered as wilderness. While there were few restrictions in these wilderness areas on activities such as logging, mining, and grazing, the administrative designations allowed the Forest Service to preserve the ability to control land development in the future by keeping lands out of the hands of the National Park Service.[16]

By the 1930s bureaucratic competition over land management had intensified. From 1933 to 1939, the National Park Service took over 1.5 million acres of the U.S. national forests.[17] In response, the Forest Service ramped up its creation of wilderness. According to David Gerard, "Park Service expansion appears to have been the central reason for Forest Service wilderness expansion."[18]

Competitive pressure was not the only motivating factor. Several conservationists within the Forest Service were also pushing for wilderness. Chief among them was Bob Marshall, one of the founders of the Wilderness Society. While Marshall proposed vast amounts of Forest Service land be converted to wilderness, the lands that were actually converted also tended to be lands in which the National Park Service was interested. While the preservationists were trying to protect lands, the Forest Service was simply trying to protect its budget. The Forest Service intended to allow these lands to be developed in the future, but a transfer to the National Park Service would have prevented that. Rather than lose the lands entirely, the Forest Service opted to "protect" them, at least in the short term. By the 1950s, they had approved non-wilderness activities in 20 percent of the areas they had designated as wilderness.

Forest Service officials were not supportive of declaring lands as permanent wilderness because having a statutory wilderness system would have eliminated the option for future development that was the Forest Service's original intention. A demand for commodity production from the nation's forests was increasing, and the Forest Service "remained an essentially utilitarian agency eager to meet that demand."[19]

Fears that the Forest Service would not protect areas it had designated as wilderness played a significant role in the development of a more robust wilderness policy. "The impetus for a wilderness act," according to Craig Allin, a political science professor at Cornell, "came more from the perception that designated wilderness was in danger than from the perception that too little had been designated."[20]

Case Study: SUWA

The Wilderness Act of 1964 created the National Wilderness Preservation System in an effort to protect lands perceived as "untrammeled by man." Prior to the bill's passage, western congressional representatives, such as Wayne Aspinall of Colorado, demanded that all wilderness be designated by an act of Congress.[21] Many western economies depended on the environment for resource extraction and commodity production. The requirement of congressional approval for future wilderness prevented federal agencies from declaring areas as wilderness unilaterally, as the Forest Service had done, and ensured that elected officials would have some say in how land would be used in their respective states.[22] Once a new act of Congress passed, designated land would become part of National Wilderness Preservation System. The act confined future designations to land that fit the following definition of wilderness:

> undeveloped Federal land retaining its primeval character and influence, without permanent improvements or human habitation, which is protected and managed so as to preserve its natural conditions and which (1) generally appears to have been affected primarily by the forces of nature, with the imprint of man's work substantially unnoticeable; (2) has outstanding opportunities for solitude or a primitive and unconfined type of recreation; (3) has at least five thousand acres of land or is of sufficient size as to make practicable its preservation and use in an unimpaired condition; and (4) may also contain ecological, geological, or other features of scientific, educational, scenic, or historical value.[23]

Since the Wilderness Act of 1964 was passed, controversies and litigation have often surrounded the creation and subsequent use of designated wilderness areas. Litigants for both preservation and extractive activities have been highly active in these controversies. One such group, from the preservation side of the issue, is the Southern Utah Wilderness Alliance (SUWA).

SUWA, a 501(c)(3) nonprofit established in 1983, has established its reputation as an uncompromising advocate for the preservation of wilderness. Its stated goal, in 1983, was "to defend America's redrock wilderness from oil and gas development, unnecessary road construction, rampant off-road vehicle use, and other threats to Utah's wilderness-quality lands."[24] Often using litigation, SUWA works toward achieving "the preservation of the outstanding wilderness at the heart of the Colorado Plateau, and the management of these lands in their natural state for the benefit of all Americans."[25]

In 2009, SUWA, along with a coalition of environmental groups including the Sierra Club, the Wasatch Mountain Club, and the Natural Resources Defense Council, supported the America's Red Rock Wilderness Act of 2009.[26] The bill stated that its goal was to "designate as wilderness certain Federal portions of the red rock canyons of the Colorado Plateau and the Great Basin

Deserts in the State of Utah for the benefit of present and future generations for people in the United States." Unfortunately, designating the land as wilderness "would lock up 9.5 million acres of land in Utah and block energy development, job creation and public land access."[27] Not one of the bill's 146 sponsors was from Utah.[28]

In fact, Illinois Senator Dick Durbin (D) and New York Congressman Maurice Hinchey (D) introduced the 2009 bill to "designate as wilderness some of our nation's most remarkable, but currently unprotected public lands."[29] With support from SUWA, the Utah Wilderness Coalition, and other environmental entrepreneurs, two D.C.-based politicians pushed for a bill that would create huge opportunity costs for local communities throughout Utah.

For example, in 2002, a U.S. Geological Survey estimate indicated that the area under consideration contained "roughly 65 million barrels of recoverable oil and 1,495 trillion cubic feet of recoverable natural gas," an amount large enough to be economically significant.[30] Furthermore, this estimate has not been updated to account for advances in technology that have increased the amount of recoverable natural resources. Political entrepreneurs in Washington and SUWA rationally continue to ignore the economic costs of the proposed America's Red Rock Wilderness Act and instead emphasize the "true" value of wilderness as untouched by human hands.

Using the Law to Keep Them off the Land

The Wilderness Act has been interpreted in such a way that many potential users are excluded. Wilderness managers must navigate conflicting interests because part of the law focuses on "use and enjoyment" and other parts emphasize "protecting . . . biophysical conditions and visitor experiences."[31] Wilderness management generally places conservation as the highest priority in regard to wilderness, even when other uses of the land may be valued. A Forest Service publication titled "Wilderness Management 101," reads, "Where a choice must be made between wilderness values and visitor or any other activity, preserving the wilderness resource is the overriding value."[32] Navigating the ambiguity of the act, wilderness management has resorted to restrictions as a means of preserving "wilderness character." Some groups, like the International Mountain Bicycling Association, support the idea of wilderness protection but disagree with blanket restrictions that exclude particular forms of recreation within wilderness boundaries. Would-be users like mountain bikers are excluded from using wilderness areas because of a "mechanical transport" provision of the law. In a *New York Times* op-ed article titled "Aw, Wilderness!" Ted Stroll writes,

> Part of the problem is that many of today's common outdoor activities were unheard of in 1964, including trail cycling and wind-powered skiing. In forbidding them, the agencies invoke the Wilderness Act's ban on "mechanical transport." But the act's legislative history makes clear that Congress never intended to stop people from using their own power to travel or shepherd their children, or from using light mechanical assistance that leaves no lasting trace.[33]

Stroll describes an evolving contradiction between preservation and access that has resulted in increasingly strict interpretations of the law that forbid "signs, baby strollers, certain climbing tools and carts that hunters use to carry game."[34] "In 1980, Congress authorized bicycling in Montana's Rattlesnake Wilderness, but the Forest Service refused to allow it."[35] Opponents of wilderness point out that the ambiguity of the act and its increasingly narrow interpretations of permitted recreation force management techniques that offer exclusive, unlimited access to one main group (hikers), while excluding other groups of users.

Environmental political entrepreneurs will use management's requirement to uphold the ultimate goal of preserving wilderness character to achieve their political ends. An example of this is the case of *SUWA v. Norton,* which directly challenged the management of wilderness study areas by the Bureau of Land Management (BLM). In this suit, SUWA alleged that the bureau "violated the Federal Land Policy and Management Act . . . and the National Environmental Policy Act . . . by not properly managing off-road vehicle and/or off-highway vehicle (collectively, ORV) use on federal lands that had been classified by the BLM as Wilderness Study Areas (WSAs) or as having 'wilderness qualities.'"[36] Although not technically designated wilderness areas, Wilderness Study Areas are regulated according to the Wilderness Act of 1964. Section 4(c) of the Wilderness Act spelled out activities that are prohibited in wilderness areas, including "no temporary road, no use of motor vehicles, motorized equipment or motorboats, no landing of aircraft, no other form of mechanical transport, and no structure or installation within any such area."[37] This suit, originally filed on October 27, 1999, contained allegations that the BLM had "'failed to perform its statutory and regulatory duties' by not preventing the harmful environmental effects associated with ORV use."[38] The Utah District Court rejected SUWA's request that nine off-road vehicle areas be closed, but the decision was reversed by the 10th Circuit Court of Appeals. After appealing to the Supreme Court, a 9–0 vote was cast in favor of the bureau because the justices found the case was "a general complaint based on policy differences."[39]

As true political entrepreneurs, SUWA did not give up on their "management through litigation" techniques.[40] SUWA filed another lawsuit in 2009 over the designation of 20,000 miles for ORV use in the Greater Canyonlands, and in 2012, they filed a lawsuit over the same issue in Richfield, Utah. SUWA's intention was to "turn this area into protected wilderness" because "doing so would eliminate any ORV recreation."[41] Although the Utah Federal District Court did not assign a wilderness designation to the Richfield area, the court "reversed BLM's off-road vehicle (ORV) trail designations because BLM failed to minimize the destructive impacts of ORV use on streams, native plants, wildlife, soils, and irreplaceable cultural sites and artifacts."[42]

Political Gridlock

Wilderness designations, according to wilderness advocates, offer individuals spiritual, ethical, and economic incentives. Jan G. Laitos and Rachel B. Gamble describe the importance of perceived ecological, environmental, and spiritual values to wilderness advocates, writing, "the mere idea of unspoiled wild land in its natural state has value to many individuals, who perceive such lands as a necessary component of human existence on this planet."[43] A late-1940s article in the Wilderness Society's *Living Wilderness* magazine claimed that wilderness "is a natural mental resource having the same basic relation to man's ultimate thought and culture as coal, timber, and other physical resources have to his material needs."[44] In his memoir, Michael Frome, an environmental journalist, states, "wilderness, above all its definitions, purposes, and uses, is sacred space, with sacred powers—the heart of a moral world."[45]

The moral obligation to maintain wilderness is often endowed with religious overtones. This began with Ralph Waldo Emerson, who wrote "about the transcendent possibilities of 'essences unchanged by man; space, the air, the river, the leaf.'"[46] Religious nature also inspired John Muir, who wrote "All the wild world is beautiful . . . everywhere and always we are in God's eternal beauty and love."[47] In the introduction to Muir's *A Thousand-Mile Walk to the Gulf*, his friend and biographer William F. Badè wrote, "Muir's love of nature was so largely a part of his religion that he naturally chose Biblical phraseology when he sought a vehicle for his feelings."[48]

Unfortunately, the quasi-religious activism of environmental entrepreneurs often leads to political gridlock because advocates refuse to back down from their positions, which are based not on science or reason, but on morality.

Wilderness benefits are not always presented from a moral perspective. In fact, quite fantastic claims are made for the economic benefits of designated wilderness. According to the Wilderness Society, "designated wilderness areas on public lands generate a range of economic benefits for individuals, communities, and the nation—among them, the attraction and retention of residents and businesses."[49] Some of the claimed economic benefits from designated wilderness fall to retailers, who enjoy around $300 billion in sales for low-impact recreation in the form of "gear, food, lodging, entertainment, and transportation" near wilderness areas.[50] In addition, "non-motorized outdoor recreation pumps $730 billion into the United States economy annually, and supports about 6.5 million jobs."[51] These claims all suffer from the same fallacies made by defenders of municipal sports facilities. Those studies claim large economic benefits from building a new facility, usually at taxpayers' expense, but the studies of actual effects show that the claimed benefits are not new spending, just a reallocation from some other activity.[52]

In order to uphold the spiritual and moral obligations to protect wilderness areas, organizations such as WildEarth Guardians are innovators who look for opportunities to advance their agenda through the use of the courts and spiritual or emotional appeals. WildEarth Guardian's value statement reads, "We believe in nature's right to exist and thrive. We act on this belief with compassion and courage by preserving the wild world."[53] WildEarth Guardians has advanced a strategy called "Litigating for Wild" to achieve their stated mission, which, according to their website, is "to confront the threats facing the beauty and diversity of the American West." Their primary strategy is to file lawsuits to uphold their interpretations of environmental laws. They also "use public awareness campaigns and political pressure to protect wildlife, wild places, and wild rivers." The group also organizes and participates in riparian area restoration projects. Their 2010 budget was just over $1.6 million, $490,000 of which was from government grants. Ten percent of their income was from settlements of legal actions. In their 2010 annual report,[54] they highlight legal actions filed to protect 100 species they believe are imperiled. They also highlight the use of a "little-known but powerful" section in the Clean Water Act to designate 700 miles of streams and rivers in New Mexico as "outstanding waters." In 2012 they announced a plan to challenge "every single new" coal lease in the Powder River Basin between Wyoming and Montana. WildEarth Guardians are typical political entrepreneurs, exploiting opportunities and capturing the profits of their alertness in the form of policy rents.

Unfortunately, designating wilderness areas comes at the expense of other users of the land. Local governments and rural groups, especially in the western United States, typically oppose additional wilderness designations. The reason for this is because wilderness designations limit traditional sources of economic growth and reduce the autonomy of local leaders. In other words, the costs of wilderness are not evenly distributed; local communities bear a disproportionate share of the costs of wilderness designations. One of the main arguments against wilderness is that it divorces land from a system of property rights in which exchanges are mutually agreed-upon between parties and the costs are confined to individuals voluntarily engaging in a particular exchange. Clint Bolick discusses this aspect of the system of wilderness in the United States, writing:

> Unlike the market system, in which buyers and sellers are responsible for the consequences of their own decisions, the political system can

be exploited to allocate costs to one group and benefits to another. For instance, when land is removed from development for recreational or aesthetic purposes, those directly receiving the benefits rarely shoulder the full costs. Instead, their wishes are subsidized by taxpayers in general, thus effectuating a "transfer payment" of sorts.[55]

Preventing Wilderness Management

Four different agencies—the Forest Service, Fish and Wildlife Service, National Park Service, and the BLM—are responsible for managing wilderness on federally owned land. Because preservation of wilderness is the highest value of the Wilderness Act, environmental groups that cling to the balance of nature ideology often oppose management activities even as basic as restoring water pipelines or cutting trees to manage insect infestations and protect recreationists from injury. In doing so, they rely on provisions of the Wilderness Act that consider the use of technology and modern management techniques as violations of a wilderness ethic. Such an ethic insists that even when human actions are beneficial ecologically, they ought to be avoided in order to preserve nature's wildness. A series of questions from Peter Landres regarding management techniques illustrates the hands-off wilderness ethic:

> The question is not whether we can take action, it is whether we should. Should we spray herbicides to control non-native invasive plants? Should we provide water to bighorn sheep that are now cut off from their seasonal sources? How about felling trees that have grown because of fire exclusion and are now ladder fuels threatening old-growth trees? Or periodically dumping lime in a stream to buffer acid deposition? Or removing landslide debris from a stream that now blocks spawning of listed fish?[56]

By evaluating the consequences of management techniques and the incentives of managers, we can understand why the Wilderness Act sometimes yields poor outcomes. If permitting violent crown fires by refusing to cut standing dead fuels, allowing debris to damage watersheds and dramatically alter fish populations, and permitting insect infestations to degrade scenic landscapes to a natural, decaying condition are acceptable, and if this is for the purpose of pursuing a higher, ethical end of preserving a hands-off approach to wilderness management, then doing nothing is an acceptable management tool. Instead, if the purpose of management is to preserve ecosystems and biodiversity, other management techniques must be considered. In other words, if doing nothing is the goal itself, then the Wilderness Act has met its objective. If the objective is to have a thriving ecosystem or economy, however, a hands-off approach may not be the best.

The dividing line of opinions can be drawn between those who favor manipulative management to restore "natural" conditions (post-aboriginal), and others encouraging nonmanipulative management to preserve wildness and untrammeled land.[57] The Wilderness Act itself offers little clarification as to how land should be managed. Landres describes the confusion: "Not restoring wilderness may allow natural conditions to further degrade, but taking action destroys the symbolic value of restraint and may influence natural conditions in unknown ways."[58]

The ambiguity regarding management techniques allows for political entrepreneurship in the form of legal opposition from environmental groups when agencies engage in activities that don't explicitly prioritize ecosystem conservation but instead preserve recreational experiences for humans or restore historical buildings located in wilderness areas. Both of these activities are protected within the Wilderness Act in Section 4(b), where it states that "wilderness areas shall be devoted to the public purposes of recreational, scenic, scientific, educational, conservation, and historical use." For instance, Wilderness Watch speaks out against rescue training operations using helicopters in La Madre Mountain Wilderness, restoration of historical mountain lookouts in Washington and Montana wilderness with helicopters and chainsaws, and the Forest Service's removal of hazardous trees in the Joyce Kilmer-Slickrock Wilderness using chainsaws and dynamite.[59] All such actions, according to Wilderness Watch and a variety of court decisions, compromise the "wilderness character" of land and violate the law.

Wilderness Watch has also expressed opposition to actions intended to preserve the biodiversity and ecological integrity of protected lands. Using "minimum requirement" language of the Wilderness Act, Wilderness Watch fights plans to use helicopters to locate and tag big horn sheep in the Sierra

Nevada range to monitor and protect the animals. They also oppose plans to install guzzlers in Arizona wilderness areas to aid declining desert bighorn and Sonoran pronghorn populations and plans to operate a fish stocking and hatchery management program in California.[60] Despite the possibility of biodiversity loss, many environmental entrepreneurs insist that preserving the concept of wilderness is paramount among competing environmental priorities.

For agencies that administer wilderness areas, doing nothing is a safe way to avoid lawsuits from environmental groups. Political entrepreneurs such as Wilderness Watch, who sue over wilderness infractions, rely heavily on the preservationist language of the Wilderness Act to defend their perception of a balanced, delicate wilderness. Frequently cited is Section 2(a), which reads, "In order to assure that an increasing population, accompanied by expanding settlement and growing mechanization, does not occupy and modify all areas within the United States and its possessions, leaving no lands designated for preservation and protection in their natural condition."[61] Environmental groups often litigate over management decisions that use too much modern machinery, but no legal standards exist for determining when an agency is doing too little to preserve the land.

The Wilderness Act causes problems in other ways. Often political entrepreneurs will advocate against a transfer of management that would benefit state and local economies. An example of this can be seen in one of the more controversial aspects of environmental law in Utah, the proposed plan of Utah's Governor Gary Herbert, which would shift federally owned lands located in Utah to state control and management. This issue began in March 2012, when Governor Herbert signed the Transfer of Public Lands Act, which demanded the transfer of over 30 million acres of federal land to state control.[62] SUWA has been opposed to this measure from the beginning, arguing that such an action would place these areas at greater risk of damage and degradation. SUWA's concerns surrounding this measure stem from what they see as the inability of the State's government to properly manage these lands. SUWA makes the assertion that "the federal government currently spends between $200 and $300 million per year managing public lands in Utah."[63] SUWA maintains that if these federal lands were to be transferred to state ownership, "Utah taxpayers would be stuck with the cost of managing them."[64]

Unexpected Consequences of Political Entrepreneurship: The Mountain Pine Beetle

The mountain pine beetle is endemic to the West. Each summer, mountain pine beetles travel to nest in trees in the Black Hills National Forest, condemning the trees to die. Beetle larvae deprive trees of essential nutrients, and a fungus living in the beetle turns the tree's trunk blue, causes dehydration, and ultimately kills the tree. Native to many forests of the West, the pine beetles' actions are part of a natural process. The rate at which the beetles are thriving and destroying other ecological components, however, has caused alarm. Some point to climate change and mild winters as fostering bark beetle reproduction and preventing die-off in winter.[65] Others point to past policy of suppressing forest fires, which means that trees of the same species, size, and age that are growing so close together have created ideal conditions for the pine beetle to flourish.[66]

The balance of nature approach to the pine beetles has far reaching effects. Hikers have limited access to trails and areas because the beetles create the danger of falling trees. Industries that depend on timber and tourism are impacted because the trees are being destroyed, and wildlife that depends on the trees for food and habitat all feel the effects of the pine beetle. Beyond losing tourism revenue because "nobody wants to see a dead forest," Carson Engelskirger, forest programs manager for the Black Hills Forest Resource Association, estimates that about 1,500 jobs directly depend on the logging industry around the Black Hills National Forest.[67] Engelskirger explains that trees are commercially viable for one year after they have been infested with beetles. After that, the tree's condition is degraded to a point at which the timber is of little value.[68] Furthermore, infested trees harvested within one year are less profitable because of their characteristic blue stain.

There is considerable debate among political entrepreneurs concerning the best practice for managing the beetle infestations. Environmentalists supporting the balance of nature ideology say forest management is doing too much, while local residents bearing the economic costs say officials haven't been doing enough to fight the beetles in South Dakota.[69] In comments before a Congressional subcommittee, Black Hills National Forest Supervisor Craig Bobzien explained that because of the "overly dense forest" the Forest Service engaged in thinning. "We thin to reduce the threat of severe wildfires to communities, improve forest health, and to improve wildlife habitat," he said.[70] Brian Brademeyer, on the other hand, alongside other members of Friends of the Norbeck, insists, "This whole beetle hysteria that dead trees are more fire-prone than green trees, it's all self-serving, logging institution-building. They should just stay in their offices."[71] Brademeyer believes that the pine beetle outbreak is a natural process that will eventually benefit the forest in terms of tree diversity and preserve an ethic of nonintervention. In Brademeyer's view, the pine beetle is simply an excuse for logging interests to access restricted areas and cut down the forest, therefore violating the state of nature in the Black Hills Forest.

Despite objections like Brademeyer's, the Forest Service management uses physical and chemical techniques to control pine beetle populations. Forest thinning, controlled burns, and planting of diverse tree species all help reduce pine beetle viability. For example, pheromone treatments in the form of pouches help trick beetles that certain trees are already infested and prevent the insects from moving in. High-value trees can also be chemically sprayed to protect the aesthetic appeal of the area such as that surrounding Mount Rushmore. Another process called "chunking" involves cutting infected trees into segments less than 24 inches long, killing the beetles before they can migrate and infect the still-living trees nearby.[72] While the Forest Service can experiment with different management techniques in some areas of the Black Hills Forest, they are prohibited from using techniques to control beetle populations in areas that are designated as wilderness, such as the Black Elk Wilderness.

The dilemma for wilderness management is that the Wilderness Act asks management to preserve and protect wilderness for future use but also stresses that wilderness must "[retain] its primeval character and influence" and must be managed so that it "appears to have been affected primarily by the forces of nature."[73] Trees falling across trails and preventing access to hikers, erosion accelerated by dead forests, and violent crown fires that are fueled by standing dead material are hardly desirable, but they are the result of the forces of nature. To complicate matters even more, the Wilderness Act speaks of preservation using unclear terms like "wilderness character" that are subject to arbitrary interpretations. In a documentary on the situation in the Black

Hills Forest, a Forest Service official noted, "If you do nothing you get something you don't expect."[74] The official goes on to say that a pristine forest is not going to naturally exist forever because random natural processes, such as pine beetle outbreaks, will brown and thin the forest. Relying on the balance of nature ideology and hoping that plots of wilderness will somehow preserve scenic landscapes, wildlife, ecosystems, recreational opportunities, and a variety of other desirable ends without management is naïve and, in many instances, produces internal and external harms felt by property owners, recreationists, and wildlife alike.

Looking at the pine beetle example further, management in designated wilderness areas is particularly challenging given restrictions on motorized equipment like chainsaws, used elsewhere to thin overgrown forest and manage beetle infestations. The impacts of the pine beetle outbreak are not localized to officially designated borders of wilderness, nor can the impacts of humans be isolated from wilderness areas. Trees infested with pine beetles can be cut, removed, thinned, sprayed, or treated with pheromones, but many of these tactics are seen as out of harmony with preserving the let-nature-be ethic of the Wilderness Act.

Conclusions

Heralded by supporters as "the best, most practical mechanism to ... preserve wilderness *in perpetuity*," the Wilderness Act gained momentum when the balance of nature ideology dominated conservation conversation.[75] Language in the law reflects an oversimplified dichotomy between man and that which is natural or wild. According to the law, land is inherently wilder when it is "affected primarily by the forces of nature, with the imprint of man's work substantially unnoticeable."[76] Such a simplistic division fails to address realities about natural environments, which do not, in fact, exist in a vacuum, independent of human impacts. If the desired goal of a wilderness area is to preserve a landscape as it existed before European colonization, quarantining land from human impacts and expecting nature to freeze in time will not work. The story of the mountain pine beetle is just one example that wilderness may need to be managed to achieve a desired end.

Craig Allin describes the Wilderness Act as "a bundle of contradictions, ambiguities, and redundancies," noting how "the statutes that create wilderness areas, and the statutes and regulations by which they are administered, are the product of competing political forces, often with incompatible objectives."[77] While the act expects management to let nature be, it also describes a variety of ends for which the land should be preserved (e.g., science, education, scenery, history). The compromises the act makes are hinged on the assumption that nature will largely balance itself to make several competing objectives possible. Once this assumption falls apart, however, the Wilderness Act fails to adequately protect the "wilderness resource" and may allow conditions that are inappropriately deemed part of nature's balancing act, to degrade further.

"When the Wilderness Act was written," writes Landres of the Aldo Leopold Research Institute, "it was probably assumed that simply not taking direct actions within wilderness would protect natural conditions."[78] Science has progressed beyond 1964, and ecologists now understand that nature must be managed, ecosystems (however defined) do not return to some equilibrium if left undisturbed, and today's ecosystems did not result from nature taking its course. Modern environmental policies, including the Wilderness Act, are based on false and misleading assumptions. Despite these false assumptions, wilderness areas continue to be designated. David N. Cole writes, "The size of the wilderness system has increased more than expected. It is already twice the 'outside maximum' of 50 million acres projected in congressional testimony by Howard Zahniser, the principal architect of the Wilderness Act (US Senate 1961)."[79] Wilderness, as defined in the Wilderness Act, is a cultural, political, and legal creation built on outdated ecological principles. Before designating new wilderness areas, our understanding and management of wilderness must be updated to accommodate scientific advancements and the concerns of local communities and individuals.

Our nation's relationship with wilderness presents a conflict of values. We are faced with competing goals—choosing a policy that values lands for the sake of their "untrammelled-ness," or a policy that values lands for their health and biodiversity. We are faced with a fight over resources, where some value the very real aesthetic and spiritual resources that lands provide, and others value the natural and energy resources. We are faced with conflicting and competing interests and actors, energy companies seeking more resources, wilderness lovers seeking a solitude and retreat from the modern world, environmental advocacy groups seeking donations, politicians seeking reelection, and federal management agencies seeking higher budgets while simultaneously avoiding being sued for managing too much. This complicated situation doesn't look like it will be resolved soon, but eliminating the harmful and useless myth of the balance of nature ideology is a good place to start.

10

Renewable Energy Legislation

IN AUGUST 2003, a massive power outage occurred in the northeastern United States. The disruption spanned the border between the United States and Canada, leaving 50 million people without electricity.[1] Eight states were affected by the power loss, making it one of the most significant blackouts in American history.[2] Because the outage occurred only two years after 9/11, authorities assumed a terrorist attack, and New York Governor George Pataki mobilized the National Guard. It was soon discovered, however, that the power loss was not the result of terrorist activity but of some overgrown trees hundreds of miles away in Ohio.[3]

People on both sides of the U.S.-Canadian border renewed their interest in energy policy when they realized that a seemingly insignificant action could have widespread and devastating consequences. The total cost of the blackout was estimated to be between $4 billion and $10 billion in the United States alone.[4] President George W. Bush partnered with the Canadian prime minister to call for the creation of a bi-national task force to "investigate ... and [find] ways to reduce the possibility of future outages,"[5] but this was not enough to satisfy the increased concerns of Congress. These concerns became the basis for the eventual passage of the Energy Policy Act of 2005.[6] The new act included over 500 pages of energy regulations ostensibly designed to modernize U.S. energy practices. It sought to diversify the energy market, while securing already functioning energy systems.[7] Regulations on electricity, renewable fuel standards, tax incentives, energy efficiency, and domestic energy production were all included in what was "the first omnibus energy legislation enacted in more than a decade."[8] Among the numerous incentive structures and policy outcomes created by the act, one in particular gained national notoriety: federal incentives for renewable energy development.

At least five separate incentives were created by the Energy Policy Act of 2005 for the purpose of promoting alternative energy projects and technologies: the Renewable Electricity Production Tax Credits, Clean Renewable Energy Bonds, Residential Renewable Energy Tax Credits, energy goals and standards for federal buildings, and last, a loan guarantee program from the Department of Energy (DOE).[9]

In this chapter, we explore how the loan guarantee program allowed political entrepreneurs to gamble taxpayer dollars on renewable energy. Specifically, we examine the political rise and economic demise of Solyndra, a solar energy company that received $500 million in federal loan guarantees under the Energy Policy Act, caused by the political entrepreneurship that occurred within the Obama White House, the DOE, and Solyndra itself.

The reader will note that this chapter is fundamentally different than some of the others. It is primarily a case study that details a more contemporary example of manipulation by crafty political entrepreneurs. Readers will recognize many aspects discussed in other chapters, including demonstrations of rent-seeking, political entrepreneurship, and costs to taxpayers.

Although support for renewable energy is not directly correlated to the flawed balance of nature theory, it arises from a pervasive implicit and sometimes explicit assumption that it is morally wrong to rely on energy that is not designated as renewable. The idea is that we should not use what we can't replace or that extraction is fundamentally an incorrect and unnatural practice.

Solyndra

In the late 2000s, the Obama Administration pushed to fund Solyndra in order to bolster the president's image as being environmentally friendly. Solyndra, a California solar panel manufacturer, was fed by the federal government's preferential treatment of the renewable energy industry. The federal government's confidence in Solyndra as a "sure winner in the solar industry" resulted in a DOE loan that amounted to $535 million and the employment of over 1,100.[10]

This confidence was misplaced. In August of 2011, less than two years after being awarded its massive government loan, Solyndra declared bankruptcy, laid off all of its employees, and was raided by FBI agents for suspected criminal activity. Even worse, American taxpayers footed the bill for this failure.[11] The Obama administration took considerable heat for its uninformed and ill-advised decision to grant Solyndra funding.[12] The controversy, described as "the first serious financial scandal of the Obama administration," was a focus of much criticism during the 2012 presidential campaign.[13]

In the beginning, Solyndra's solar technology seemed promising. CEO Chris Gronet founded the company in 2005 based on an idea he had for reducing the costs of installing solar panels. Most companies manufacturing conventional solar technology create flat panels that are then installed on rooftops of homes or other buildings. Gronet's unique idea relied on covering roofs with cylindrical thin-film solar cells rather than flat panels. These cylinders encase thin-film solar cells made of copper-indium-gallium-selenide. The glass in the cylinder both seals out moisture and acts as a "sunlight concentrator" that funnels photons onto the film, producing more electricity.[14]

Gronet explained the benefits of using this new shape instead of the traditional flat panel system: "With a cylinder, we are collecting light from all angles, even collecting diffuse light."[15] Solyndra's technology, unlike conventional solar-power installations, didn't have to be tilted or "anchored on the roof with ballasts and adhesives."[16] This reduced the cost of installation and the time that it took to mount the cylinders on a rooftop. Whereas the panels have gaps that allow wind to pass through freely, Solyndra could make its solar "arrays" (a collection of cylinders) "wind-resistant, capable of withstanding winds up to 130 miles per hour."[17] A test installation in Florida survived Tropical Storm Fay in 2008.[18] The technology being developed was innovative, a fact used as leverage by political entrepreneurs at Solyndra to eventually secure federal government funding.

Solyndra's business plan also seemed to be cost-effective at the time. The company's technology, according to Gronet, would provide a means through which solar power could be harnessed at about the same cost as electricity from coal-powered plants.[19] The price of electricity generated from coal is generally around 6 cents per kilowatt-hour, while conventional solar panels average around 25 to 50 cents per kilowatt-hour. Solyndra's cylindrical design reduced costs by making its solar cells cheaper and faster to install than conventional solar panels.[20] With a unique new technology that promised to cut costs, Solyndra appeared to have a plan for success.

Under normal market conditions, Solyndra's worthiness as an investment would have been determined by its ability not only to produce a needed product, but also to do so profitably. Solyndra was unable to meet these criteria, but rather than go back to the drawing board to create more competitive innovations or business strategies, the company continued its economically unviable operations, thanks to the simultaneous convergence of three groups of political entrepreneurs. Solyndra sought government funding to continue operations, and the DOE and the Obama administration granted that funding to garner public approval through environmentally friendly policies.

Solyndra's Path to Funding

Solyndra was funded under the Loan Guarantee Program in section 1703 of the Energy Policy Act. It authorized the DOE to provide loan guarantees to finance clean energy projects that "avoid, reduce or sequester air pollutants" and "employ new or significantly improved technologies as compared to commercial technologies in service in the United States at the time the guarantee is issued." The act listed ten categories of projects eligible for funding, and Solyndra's solar technology fit nicely into the "renewable energy systems" category.

Solyndra's original application was submitted to the DOE's Loan Programs Office in December of 2006, along with applications for 143 other projects. The Solyndra application was finally considered complete in August 2008. It included financial statements, the company's business plan, and market reports for Solyndra's solar products. The total project cost listed was $713 million, with $535 million of the total cost to be provided for by the DOE loan guarantee.

As the review of the Solyndra application began in May 2008, concerns about the program were already being voiced. Members of Congress and the Government Accountability Office expressed concerns about the management of the program, specifically criticizing the DOE's "inconsistent treatment of loan guarantee applicants in the due diligence process" and its policy of making conditional commitments for funding before the required independent reports were received. Solyndra was one of those projects for which a conditional promise for funding was made before the necessary information was gathered and reviewed.

The Loan Programs Office hoped to be able to present the application to the Credit Committee and Credit Review Board by January 2009; however, this did not seem likely to happen as several key documents were still missing as late as December 2008. In addition, a credit review of the company identified potential problems with the application. In the final days of the Bush Administration, on January 9, 2009, a Credit Committee was convened, which concluded that recommending Solyndra for approval was "premature at this time" due to significant unresolved questions. A few days later, Chairman Lachlan Seward confirmed the committee's decision not to engage in further discussions with Solyndra at that time.[21] Solyndra's quest for government funding had hit a dead end.

A change of administration proved to be just what Solyndra needed to speed things along. Although President Obama described the Solyndra funding process as "a loan guarantee program that predates me" at an October 6, 2011, press conference, Solyndra's application actually was not approved under the Energy Policy Act of 2005 or during the Bush administration.[22] Solyndra's loan approval was made possible through amendments made to the Energy Policy Act by the American Recovery and Reinvestment Act of 2009, better known as the stimulus package. The Recovery Act amended the 2005 Energy Policy Act, creating section 1705, which provided additional funding for "commercially available technologies" and made the new loans more attractive by removing the applicants' responsibility to pay credit subsidy fees.[23]

When the Obama administration took office in 2009, three years had passed since the beginning of the Energy Policy Act of 2005 loan program, and the DOE still hadn't approved a single loan guarantee. This may be due in part to the fact that the Bush administration believed passage of the act was sufficient to satisfy public opinion, making active pursuit of energy developments less urgent. The Obama administration, acting as political entrepreneurs, was in a position to directly capitalize on the perceived public benefits of pursuing an environmentally friendly agenda. The administration was eager to get the ball rolling on approving funding for clean energy projects like Solyndra.

In 2009, Steven Chu, a Nobel Prize-winning physicist and former head of the Lawrence Berkeley National Laboratory, was appointed secretary of the DOE. Upon taking charge of the agency, Chu faced extreme pressure from the administration to speed up the loan process for clean energy projects. Bending to political pressures, Chu promised to do so, stating, "We're trying to streamline it so that the period of time will be reduced from a scale of 4 years to several months."[24] The additional $6 billion in funding provided by the stimulus package required that projects begin construction by September 30, 2011, and Chu wanted to make sure that this money was spent on clean energy projects before this deadline. The passage of the stimulus legislation, with its increase of clean energy funding and tight deadlines, made the Solyndra application "a top Obama administration priority."[25]

In January 2009, Solyndra's loan application was once again opened for review and potential approval by the Loan Programs Office. Throughout February of that same year, the DOE and Solyndra discussed terms of the application that continued to be cause for concern. These concerns included Solyndra's proposal of an 80 to 20 debt-to-equity ratio, which the department thought did not require the company to raise enough equity. Because the stimulus now provided funding for credit subsidy costs, department officials, including chief financial officer Steve Isakowitz, expressed his fear that if such a low ratio of equity were allowed, the new process would not provide significant incentives for applicants like Solyndra to have some "skin in the game."[26] Isakowitz worried that potential loan applicants would not be sufficiently encouraged to take responsibility for minimizing risk if their loans were to be approved and if the equity requirement was set at such a low standard.

After less than two months of negotiations on loan terms, the DOE's Credit Committee approved the Solyndra application unanimously. The agreement was subject to the condition that the Loan Programs Office provide responses to eleven "follow-up concerns" raised by an independent report conducted by R. W. Beck. These included a lack of competitor information in the report's market analysis and misalignment between the financials of the parent company, Solyndra, Inc., and those of Solyndra Fab 2, LLC.[27] Just days later on March 17, 2009, the Credit Review Board likewise approved Solyndra's $535 million loan unanimously, despite the fact that only one of the Credit Committee's follow-up concerns had been addressed.[28]

Solyndra and the DOE spent the next several months taking the necessary steps to secure the closing of the loan and working through various complications. One such complication occurred when it was revealed that the department did not consult with the secretary of the Treasury before or during the review of Solyndra's loan application. This step was required under the Energy Policy Act of 2005. Because the department waited until the last minute to consult with the Treasury officials, the consultation ended up being rushed. The department did not contact the Treasury about Solyndra's final term sheet until two days before the Credit Committee meeting and just a week before the Credit Review Board meeting.[29] One of the principal concerns of Treasury staff included the debt-to-equity split. "This is the first deal out the door and I am worried that it will set a standard for subsequent deals," said a Treasury official who further expressed his concern, noting, "this should have been 65% debt and 35% equity instead of 73% debt and 27% equity," as agreed to by the DOE and Solyndra.[30]

In response to such concerns, Treasury officials negotiated a two-day extension in order to conduct a review of the Solyndra term sheet, which laid out the terms and conditions of the loan guarantee. But political pressures got the best of the Department of Energy, and its staff ramped up the pressure on the Treasury, resulting in an expedited process and completion in just one day.[31] The review resulted in additional concerns being raised by Treasury officials, including, "the base equity commitment, restrictions on the sponsor's use of equity, overrun project costs and contingency funds, and rights to intellectual property."[32] Although the Treasury was allowed to review the Solyndra term sheet, this action was taken more to comply with technicalities than to ensure that concerns with Solyndra's loan were identified and solved. The DOE, under time constraints from the Obama administration, pressured the Treasury for a quick turnaround on the application review, resulting in red flags being ignored.

"The Solyndra Failure," an August 2, 2012, report by the House of Representatives Energy and Commerce Committee, included documents indicating that President Obama's tentative trip to California was a major consideration in the DOE's decision to accelerate the process.[33] According to the report, "the scheduling of President Obama's speech on March 19 set the timetable for the Credit Committee and Credit Review Board meetings for Solyndra."[34] The fact that the administration's desire to gain publicity by announcing a new clean energy project was given preference over the critical review process for Solyndra's multimillion-dollar loan highlights the political nature of the process. The report further found that the White House and the DOE pressured the Office of Management and Budget to speed up its review of Solyndra's loan in time for a September 4 closing event at which the vice president was to appear by satellite.[35] Citing these time constraints imposed by the White House, the report claims that "this pressure may have had a tangible impact on the Solyndra credit subsidy cost calculations as OMB staff stated that they did not have time to adjust the factors in their modeling due to the time constraints."[36]

In the end Solyndra's efforts to raise equity were successful. By July 2009, the company had raised $280 million in equity with the help of Goldman, Sachs & Company, of which $198 million was to be applied to Solyndra's DOE loan guarantee project. Argonaut, the investment arm of the George Kaiser Family Foundation, became the largest shareholder, contributing $130 million.[37] George Kaiser is an Oklahoma billionaire and was a key fundraiser for Obama, especially during the 2008 presidential campaign.[38] According to the House Committee on Energy and Commerce report, Argonaut ultimately held a 39 percent stake in the company, and George Kaiser himself was closely involved in approving and disapproving a variety of political tactics being proposed.[39]

After raising sufficient equity, Solyndra Fab 2, LLC, was given a rating of "BB–," by Fitch Ratings, indicating a "speculative" investment score. According to the company, this meant that Solyndra's loan guarantee indicated "an elevated vulnerability to default risk, particularly in the event of adverse changes in business or economic conditions over time."[40] Reasons given for this rating included the extreme competitiveness of the solar market.[41]

As late as September 1, 2009, just three days before the White House publicity event was scheduled, the Office of Management and Budget was still debating whether the DOE had provided enough information to address all remaining concerns with the Solyndra loan application. In particular, the office voiced concerns about the Solyndra product's efficiency and whether it would be able to compete with traditional solar technology. The office wanted to see a comparison showing how Solyndra's panels compared to conventional solar panel installations in actual production of electricity, but only modeling estimates were provided instead of real data.[42]

Despite White House concerns that the Office of Management and Budget would not complete its review on time, the Solyndra application was given final approval. On September 4, 2009, a groundbreaking event was held in Fremont, California, with Vice President Joe Biden appearing by satellite and Secretary Chu appearing in person.[43] The groundbreaking began with an announcement that the loan guarantee had finally been closed. California Governor Arnold Schwarzenegger attended the event and declared, "Hasta la vista to global warming," calling Solyndra a great example of both job creation and environmental protection. Throughout the event, the emphasis remained on job creation. The vice president expressed his excitement that the new project would create, "jobs that won't be exported," while reducing carbon emissions at the same time.[44]

Secretary Chu said he hoped that Solyndra's groundbreaking would "kick off" many other alternative energy projects under the loan guarantee program. The event concluded with a photo-op in which political entrepreneurs Chu, Schwarzenegger, Gronet, and Bob Wasserman (the mayor of Fremont) all took up shovels and made the groundbreaking official.[45] Solyndra's first loan from the U.S. government was now official and public. The company now had the means to begin building its new manufacturing plant, and the policymakers behind the scenes received the political capital they had been seeking.

Solyndra's attempts to get financing from "the Bank of Washington," as then-CEO Christian Gronet phrased it, were just beginning.[46] Solyndra applied for a second loan guarantee in the amount of $469 million in January 2010, but this application was ultimately denied due to concerns about excessive risk and the company's already deteriorating financial situation. It seemed that this ill-advised attempt to pick favorites was producing poor outcomes, and the government began cautiously looking at the ramifications of Solyndra's failure. According to Damon LaVera, a spokesman for the DOE, the second loan application for Solyndra didn't go far before the department and the company agreed that it should not receive further consideration, especially since the first factory was yet to be completed.[47]

Although it was just months after securing its first government loan, Solyndra was already headed for trouble. In February 2011, it became clear that Solyndra was floundering and might default on its loan, so the Obama administration helped the company refinance its $535 million federal loan and extended an additional $67 million as a lifeline in the hopes that this would save the massive federal investment from sinking. The administration took considerable heat for this decision. As an extension of the Obama administration, political entrepreneurs at the DOE were accused of pushing for this restructuring, despite warnings from federal staff in the Office of Management and Budget that the restructuring would end up costing taxpayers $168 million more than letting it fail.[48]

Solyndra's Failure

Given the significant business flaws, ignored financial concerns and the clear political agenda that surrounded the Solyndra loan application, the company declaration of bankruptcy only two years after the approval of its original loan was not surprising. In September 2011, Solyndra sought Chapter 11 bankruptcy protection and fired its over 1,000 employees. In July 2012, the company filed a Chapter 11 reorganization plan through which its unsecured claims are expected to recover between 2.5 percent and 6 percent.[49] Although private equity firms are expected to get back at least half of the $70 million or so they invested in Solyndra in early 2011 to help save the company from its downward spiral, U.S. taxpayers will be the ones that lose out, with only $24 million of the original half billion-dollar loan being potentially repaid.[50]

The knowledge necessary to accurately select the right technologies or innovations is not available to anyone. No one can foresee future market conditions or adequately inventory every possible alternative to decide which technology or company to bet on. In private markets, risks are generally spread across many investors who have incentive to make good investments. The government, on the other hand, uses other people's money and has no such incentive to invest well.

Solyndra's federal loan guarantee was the result of a highly politicized process. The Obama Administration rushed to approve the Solyndra application in order to strengthen its image as being environmentally friendly. The administration felt pressure to deliver on campaign promises regarding support for clean energy and to make this happen quickly. When it became clear that the investment was going to fail, the administration further entangled itself in the political controversy by pushing Solyndra to delay its announcement that it would lay off workers until after the November 2010 midterm elections in which Democratic control of Congress was at stake. The company complied with the administration's request and released the layoff announcement on November 3, 2010, one day after the election.[51]

Conclusions

The renewable energy incentive program described here provided political entrepreneurs with opportunities to engage in rent seeking, manipulation, and political posturing. Such programs favor specific industries by allowing policymakers to gamble with taxpayer dollars in attempts to be smarter than markets about the future.

The Obama administration promoted highly visible "green" projects in an attempt to increase voter approval—the promotion was based on neither an economically viable technology nor a savvy business plan. Public funds boosted Solyndra into initial financial success, and the Obama administration into initial political and ideological success. But even all the public funds invested could not save Solyndra from market forces, and it was soon revealed as the problematic company the market itself had already rejected. The case study of Solyndra and the Energy Policy Act is not precisely similar to other environmental legislation examined in this book. The Energy Policy Act did not emerge from the balance of nature ideology, and it does not seek to preserve huge tracts of land in a pristine state—nor is Solyndra a doomed species of owl. However, this case study does bear resemblance to the others in this book because Solyndra's rise and fall was predicated on the ecology of politics—rent-seeking behaviors and political ecology—not rational decision-making.

The agendas of the Obama White House, the DOE, and Solyndra executives converged under the umbrella provided by the Energy Policy Act.

White House operatives sought a green symbol to show the voting public. Department bureaucrats wanted to fund high-profile projects regardless of market signals. Solyndra officials were pleased to get the funding they could not get from private investors. All the political actors got what they needed, at least in the short run. The only real losers were the taxpayers, and those costs were spread so widely that almost no one even noticed.

11

Conclusions

WE'VE DEMONSTRATED THAT our environmental laws have failed us. Stopping logging operations in the Pacific Northwest won't return forests to a natural state. Saving endangered species often makes little environmental sense. Managing lands for the sake of their permanent preservation can destroy those lands. Ecosystems do not exist in steady-state equilibrium where, if left alone, they return to an idealized condition. Our most important environmental laws increase costs and delays, fail to meet their stated environmental goals, and can often lead to worse environmental outcomes.

Identifying failures and analyzing their root causes is relatively easy compared to fixing those failures and their causes. A useful way of looking at why something fails is to look at it how it was designed. Design analysis gives insight into the reasons why things do not work as expected. Every day, poor design leads us to push on a door we are supposed to pull or turn on the incorrect burner on the stove. But poor design can have even greater consequences, for example, when the Clean Water Act results in dirtier water.

A phrase often invoked among designers is, "Every system is perfectly designed to get the results it gets." The originator of the phrase is a medical doctor who was talking about health care systems. If we apply his assertion to environmental legislation and regulation, we are left with the conclusion that our environmental laws fail because they are designed to fail. That assertion sounds harsh and even suggests intentionality, but harsh statements may be necessary to ameliorate problems. Replacing warm-hearted sentimentality about nature with hard and clear-headed analysis will allow for clear direction about the possibilities for change.

Some of what we consider to be failures in environmental legislation and regulation is actually intentionally designed, for example, restrictions on private landowners or on uses of public lands. What we consider to be failures are viewed as successes by others—economic costs of wilderness or public lands management on local communities, for example. Designers of legislation and regulation, however, did not intend other failures such as air or water getting dirtier or not as clean as they would be without the legislation. Those failures simply demonstrate the disconnection between good intentions and good outcomes.

Despite the shortcomings and failures of existing regulation, we believe there are some basic principles that, if followed, can help redesign and incentivize institutions to perform as intended. Following them will alter existing management systems with more pragmatic, effective, intellectually honest policy. Given our early, extended discussion of environmental religion, myths, and outdated assumptions, we might have offered some moral or religious principles as well.[1] We do not offer any because we believe biological integrity is a practical matter. If others wish to make it a moral issue, there is room in the processes we envision for moral and religious values.

The principles we offer are:

Biological Principles

- Managing nature protects biological integrity better than does natural regulation.
- *Natural, wilderness, preservation*, and *ecosystem* are human constructs, not scientific ones.

Political Principles

- Powerful political forces are invested in existing legislation and regulation.
- Making political changes will require extensive and intensive political entrepreneurship.
- Marginal changes are more possible than wholesale changes.

- Decentralizing environmental regulation is more effective than centralizing it. That is, fifty competing answers are better than one, especially since no one knows which is the right answer.

A Way Forward

In 1995 Utah Governor Michael O. Leavitt proposed reclassifying 1.8 million acres of wilderness study areas in Utah, as designated by the Bureau of Land Management (BLM), as true wilderness. The proposal came after seventeen years of study and debate among varying interests. Despite so much work by so many people, a bill that would have designated the 1.8 million acres as wilderness failed in the 104th Congress. The failure is instructive.

The Leavitt administration, in concert with Republicans in the congressional delegation, took a hard stance against any further expansion of wilderness designation in the areas of the state the bill affected. The Utah Wilderness Association had been arguing for about 3 million acres, while the Southern Utah Wilderness Alliance (SUWA) had been working with New York Congressman Maurice Hinchey to get 5.7 million acres designated as wilderness. Dick Carter, the head of the Utah Wilderness Association, tried to mediate between the various positions and was condemned within the environmental community. At the time, Carter was a respected voice and could have been the catalyst for a winning coalition. We believe Leavitt should have embraced him. As it was, Leavitt held firm at 1.8 million acres, SUWA pushed for 5.6 million acres, and Carter was marginalized. Had Leavitt embraced Carter and his proposal, SUWA would have been marginalized.

Adopting Carter's proposal for 3 million acres would have required compromising the hard position Leavitt and his staff had taken. But, because 3 million acres was a proposal made by a well-respected voice in the environmental community, it would have been supported by a much broader coalition in Utah and in Congress. The Leavitt administration assumed it had the votes in Congress and did not need to compromise. He was wrong. Now, twenty years later, no new wilderness bill in Utah has been proposed, and suggestions from wilderness interests have increased to 5.7 million acres.

A contrast to the Leavitt wilderness proposal is the successful passage of the Washington County Growth and Conservation Act of 2008. It was a close collaboration among the various interests at work in the county—developers, environmentalists, county commissioners, ranchers, state and federal agencies, and politicians. Based on a similar effort in Nevada, the Washington County bill established wilderness areas within the county and authorized portions of federally owned land to be sold to private parties to accommodate development.

The bill was ambitious. Its important features included designating 256,338 acres of specified public lands as wilderness and as components of the National Wilderness Preservation System, which increased the percentage of wilderness acreage in Washington County from 3.5 percent to 20.5 percent; designating 165.5 miles of the Virgin River as a protected wet wilderness area under the Wild and Scenic Rivers Act, the first of such designations in Utah; authorizing the BLM to sell 5,000 acres of excess BLM lands in the county that are not considered to be environmentally sensitive and to use the proceeds to buy other sensitive lands in the county; creating two new national conservation areas—Red Cliffs National Conservation Area and Beaver Dam Wash National Conservation Area— to provide permanent protection for the endangered desert tortoise and other at-risk species in the area; establishing provisions regarding the management of priority biological areas in Washington County and authorizing grants and cooperative agreements for the carrying out of initiatives related to the restoration or conservation of such areas; releasing some wilderness study areas in Washington County from further study for designation as wilderness; designating specified federal land in Zion National Park as a component of the National Wilderness Preservation System, declaring it the Zion Wilderness; requiring the BLM to develop a comprehensive Washington County travel management plan to accommodate off-highway vehicle use; directing the secretary of the interior to designate a trail to be known as the "High Desert Off-Highway Vehicle Trail"; and providing for the conveyance of certain public land in Washington County. The bill ended years of conflict and created some certainty about federal land management in Washington County.

Representative Rob Bishop (R-Utah) built on the model used in Washington County and proposed changes in other Utah counties. Although the bill has not passed as this book goes to press, the same principles have been used, encouraging success. Bishop has sought out environmental groups in Utah to put together what he calls a "grand bargain" on public lands and is trying to create the compromise at the local level, rather than attempting to legislate from Washington, D.C. The first agreement is in Daggett County. Dagget County has agreed to support creating 80,000 acres of wilderness, a 30,000-acre conservation area, and wild and scenic designation for a portion of the Green River. In exchange, a coalition of environmental groups agreed to support conveying more than 10,000 acres from the BLM to Utah's School and Institutional Trust Lands Administration for potential energy development and a resort. The agreement also gives Daggett County rights to open travel routes on BLM land that is outside the proposed wilderness and conservation areas. The Daggett County agreement and others like it will be included in Bishop's public lands initiative bill to be introduced in 2016. Future readers of this book will be able to judge the success of the localized, incremental approach to solving public lands disputes.

Representative Bishop's initiative recognizes the need for active management of many of the public lands in Daggett County, it recognizes the symbolic power of designating wilderness, it organizes powerful political forces to work to a mutually agreeable outcome, it is an example of untiring entrepreneurship by Representative Bishop, it makes marginal changes without violating the core of any environmental bill, and it is a decentralized approach. Getting the whole grand bargain through Congress will require exceptional entrepreneurial efforts by all the affected parties, especially Representative Bishop.

The Washington County Growth and Conservation Act and Representative Bishop's efforts suggest that positive change is actually possible, but only with a great deal of political entrepreneurship, willingness to decentralize some management currently done from Washington, D.C., and recognition that public lands need to be managed. In what follows we identify some marginal changes to the environmental legislation we analyzed in the previous chapters. Each is meant to be a marginal rather than wholesale change and is just a cursory description of possibilities for positive political entrepreneurship. The description is meant to be suggestive, rather than an exhaustive exploration of the ideas.

The Clean Water Act

Many of the problems we identified in our analysis of the Clean Water Act could be reduced by the simple, yet controversial step of restricting the definition of "waters of the United States" to waters that are, in fact, navigable. Each of the states would then have jurisdiction over all other waters. We recognize that this is controversial, especially since as we write this, the Environmental Protection Agency is proposing to expand the definition of "waters of the United States." There is clearly a fundamental ideological divide between our proposal and theirs. But, we believe that the failures of the Clean Water Act justify our position. State regulators are more likely to recognize local differences and wants. As such, we believe they are less likely to engage in bureaucratic excess to the level of federal regulators. We note, however, that this is precisely the reason many environmental groups oppose decentralizing the regulatory environment.

Another seemingly small change to the Clean Water Act would be to include economic and social impacts as part of the decision process. Making that change recognizes the human dimensions of environmental policy along with the biological dimensions. Recall the road project alternative that the Clean Water Act considered feasible, even though it would require demolishing over 250 homes and fracturing a community. The requirement to include the potential impacts of great economic and social changes in the decision-making would help in better decision-making and would require regulators to more broadly consider the implications of their recommendations.

National Environmental Policy Act

The most simple change to the National Environmental Policy Act is to require the Corps of Engineers to accept the environmental assessments and environmental impact statements that are prepared for specific projects that affect waters of the United States rather than requiring completely new studies. Currently, the Corps can ignore those studies entirely and demand new ones. A single document should be sufficient for the Corps and all other agencies.

A second change to the act would be to limit the size of its required documents to something that an intelligent layperson would be able to read and understand. Currently, such documents can run to thousands of pages, which makes it impossible for interested members of the public to read and allows for a great deal of obfuscation. More analysis and more pages do not always equate to better decision-making.

Endangered Species Act

The Endangered Species Act ought to be amended to include habitat in the species equation. We mean something more than the obvious point that protecting more habitat is preferable to protecting less. We mean that before public monies are spent on protecting a particular species, there should be an assessment of the appropriate and available habitat. Such an approach would say it makes sense to spend money to rescue the whooping crane because there is appropriate and possibly adequate summer and winter habitat, and within that habitat are found the species on which the crane preys. It would *not* make sense, however, to make heroic efforts to save the California condor in the wild because the requirements of its habitat are likely never to be met again. It would also likely reject efforts to protect the lynx or wolverine in the northern Rockies since they were historically rare in that region precisely because the habitat is not well suited to them.

The single most important step that could be taken with the Endangered Species Act would be to delete "adverse habitat modification" from the regulations governing private land. This would remove the regulatory stick that encourages landowners to destroy potential endangered species habitat rather than face the restrictions on their management practices that the law would require. In place of the stick of regulation, it might be possible to offer carrots—incentives for preserving and even creating habitat. One possibility is conservation rental contracts.[2] Such contracts could be especially important in the southeastern United States, where private individuals own about two-thirds of the commercial forest. Attempting to put together large-scale programs under that ownership pattern is difficult, but the situation is amenable to targeted rental efforts. Funding for these contracts or other positive rewards such as bounties for attracting endangered species to your property might be funded out of a new biodiversity trust that would be funded from public land user fees.[3] Alternatively, property owners could receive tax credits for habitat maintenance or improvement.

If endangered species proponents produced rewards for landowners instead of penalties, new political dynamics would emerge. Federal agents calling on landowners in Utah's Garfield and Iron counties who harbor Utah prairie dogs would be viewed as potential friends rather than enemies. Habitat protection would proceed on a local level rather than under edicts from Washington, D.C.

Wilderness Act

Since several million acres of land is held in a regulatory and use limbo because they are wilderness study areas, we suggest sunset provisions on creating wilderness from those lands. If the study areas are not recommended for wilderness designation within, say, ten years, they revert to multiple use management rather than as de facto wilderness, which is how they are now managed. The Washington County Growth and Conservation Act and Daggett County agreement described above are examples of how to move forward; deadlines in the form of sunset provisions could force many more such agreements.

Clean Air Act

The Clean Air Act could be amended to require whole economy modeling when the Environmental Protection Agency proposes a new rule. A whole economy model takes into account the interconnected and cascading effects of a proposed rule across an entire market. Such modeling produces very different results about job gains or losses than the current jobs impact formulas used by the agency.

Something we did not address in our chapter on the Clean Air Act is the secret science used by the Environmental Protection Agency to justify rule changes. EPA commissions studies and allows those conducting the studies to keep the data, modeling, and protocols secret from the public. Only the results are released. As a result, there can be no independent replication or verification of the study and EPA is able to make unverifiable claims. Such a policy flies in the face of the scientific method and the foundations of an open society. The solution is simple; free the data so it is part of a public, not a secret record.

As with the Clean Water Act, we believe the Clean Air Act should include economic and social impacts as part of its decision processes. Making that change recognizes the human dimensions of environmental policy along with the biological dimensions. It will require regulators to more broadly consider the implications of their recommendations.

Energy Policy Act

A rather simple yet powerful change to the Energy Policy Act would be to replace the subsidies it creates with prizes for achieving defined improvements to alternative energy systems. Although a system of prizes could still distort energy research and development, it would at least encourage innovation over rent seeking. There are many historical examples of such prizes, the most well known being the Longitude Prize offered by the British government in 1714. Prizes could be established not for particular methods, such as improvements in photovoltaic cells, but for demonstrated improvements in outcomes regardless of the method.

A second change to the Energy Policy Act would be to repeal the ethanol Renewable Fuel Standard. This proposal has been considered for years but has always been defeated by farming interests and politicians from those states. That coalition's power is declining, however, as data about the environmental and human costs of producing gasoline from corn and other crops have become more available. There would still be uses for ethanol in the fuel mix, but those uses would be determined by market demands, not regulations.

The Importance of Moving Forward

We were recently involved in a project that proposed a new, three-mile road from an interstate highway to a municipal airport. The project required compliance with nearly every major environmental law in the United States. Although the construction of this road was expected to take only two years, permitting and analysis alone required nearly eight years and cost millions of dollars. Each individual law incrementally added to the delays and costs of the project. The skilled hands of political entrepreneurs carefully manipulated each law. Delays were maximized, the deep pockets of the transportation agency were emptied, and mitigation was demanded. Each agency, with its political entrepreneurs, was able to extract rents from the project, usually in the form of imposing its own, often idiosyncratic values and rules. When the next road is proposed, the same groups will use this project to show what must get done to get buy-off for a new road.

Our example is relatively minor compared to the proposed Keystone XL Pipeline. In September 2008, energy company TransCanada applied for a cross-border permit from the U.S. State Department to begin construction of the Keystone XL Pipeline from the U.S.-Canada border to the Gulf of Mexico. This scenario is prime habitat for political entrepreneurs as it proposes to build an oil pipeline across multiple states, federal lands, U.S. waterways and wetlands, and endangered species habitat. The proposal triggered protests and requirements to comply with practically every federal legislative act discussed in this book.

After three years, the State Department finished its first final environmental review of the proposed extension and denied it. TransCanada reapplied for a permit with an altered route. Once again reacting to the politics over TransCanada's application, President Obama stated that he wouldn't approve the permit unless the "project does not significantly exacerbate the problem of carbon pollution."[4]

In January 2014, the State Department released its second final environmental impact statement for the Keystone project, coming to the same conclusion as the first: that the project will have an insignificant effect on greenhouse gas emissions. Despite having gone through the review process twice, with the same conclusions reached each time, President Obama decided to kill the project.

Two environmental impact statements under the National Environmental Policy Act have been completed since 2008, the BLM has considered different resource plans under Section 404 of the Clean Water Act and Section 309 of the Clean Air Act, and states have been consulted under the Energy Policy Act. The Fish and Wildlife Service found that the project would have been likely to adversely affect the endangered American burying beetle (*Nicrophorus americanus*). This discovery would have required "conservation measures and compensatory mitigation, such as trapping and relocating beetles, special lighting restrictions (the beetles are attracted to light), and establishment of a habitat conservation trust."[5] In addition to potential risks to endangered species, violating the Clean Water Act was a possibility because oil spills could contaminate water, specifically groundwater.

Delaying the Keystone project on the grounds that it will cause oil spills will have the unintended consequence of causing more oil to be spilled, not less. That's because railcars will continue to move the oil instead of the Keystone pipeline, and railcars are likely to spill more than double the amount that the pipeline would. In other words, political entrepreneurs who fought the Keystone XL project using the laws described in this book have ensured that more oil spills will occur rather than fewer.[6] The outcome is unintended and ironic, but expectable.

Some of our own understanding of how to improve on the regulatory approach embodied in the foundational environmental laws comes from the classic 1934 essay by Aldo Leopold titled "Conservation Economics."[7] He began the essay by noting that the accepted theory of his day about the birth of the moon was that a large planet passed near enough to pull a large piece of the earth into space as a new heavenly body. He compared the birth of conservation programs to that same process.

> Conservation, I think, was "born" in somewhat the same manner in the year A.D. 1933. A mighty force, consisting of pent-up desires and frustrated dreams of two generations of conservationists, passed near the national moneybags whilst opened wide for post-depression relief. Something large and heavy was lifted off and hurled forth into the galaxy of the alphabets. It is still moving too fast for us to be sure how big it is, or what cosmic forces draw rein on its career . . .
>
> [Conservation's] history in America may be compressed into two sentences: We tried to get conservation by buying land, by subsidizing desirable changes in land use, and by passing restrictive laws. The last method largely failed; the other two have produced some small samples of success.

The "New Deal" expenditures are the natural consequence of this experience. Public ownership or subsidy having given us the only taste of conservation we have ever enjoyed, the public money-bags being open, and private land being a drug on the market, we have suddenly decided to buy us a real mouthful, if not indeed, a square meal. Is this good logic? Will we get a square meal? These are the questions of the hour.[8]

"Passing restrictive laws," a method Leopold considered to have "largely failed," is what the United States adopted in the 1970s. To continue his analogy, conservation was hurled into a higher orbit with even greater infusions of government cash and regulations. To the "galaxy of the alphabets" were added FLPMA, CAA, NEPA, CWA, ESA, and many others. We believe these restrictive laws have largely failed.

We have highlighted failures that make little ecological sense but make a lot of political sense. It may be time to make the marginal changes to those laws that, in Leopold's words, "make good logic."

Appendix
Federal Land Policy

IN THE 1930S, an oil company leased federal land in east-central Wyoming and built a residence for oil workers. The house was sold to a private individual, and the house was then sold again, thirty-seven years later to Gerald Chaffin. Chaffin bought the house from the oil company in 1970 for $500, and for the next ten years, he, his wife, and family of four invested $15,000 in improvements to the house. The family was unaware that the land on which the house sat belonged to the federal government.

When the Federal Land Policy Management Act was passed in 1976, it declared that all public lands in existence then would remain in public ownership. Because it sat on land managed by the Bureau of Land Management (BLM), under the new law, Chaffin's house was considered a "trespassing" structure. Once Chaffin was notified of this, Chaffin made multiple appeals to the BLM offices in Cheyenne, Denver, and Washington. His own lawyers repeatedly told him that he had no case.

Facing an order to remove the house or serve an 18-month prison sentence and pay a $6,000 fine, Chaffin resorted to extreme measures. He didn't have the money to pay to move the house or demolish it, so the family doused their home of ten years in gasoline—along with $15,000 in home improvements—and watched it burn to the ground.[1]

As this story illustrates, federal land policy is controversial. It pits local and often private interests against representatives of federal agencies who have been delegated authority to manage those lands. Modern environmental groups often enter arguments on the side of the federal managers, but they also bring suit against the managers when the latter take actions contrary to the wishes of the environmental groups. We offer this appendix on federal land policy to help readers develop a framework for understanding the history and context of current policies.

A Storied Past: The History of Public Land Management

Prior to the late 1800s, the U.S. government's policy for public land was to dispose of it to states and individuals.[2] Following the Revolutionary War and the signing of the Treaty of Paris, the budding republic had large amounts of public land in its hands. Some was used as in-kind payments for Revolutionary War soldiers, but the states were divided over how best to allocate the land, and Maryland even refused to ratify the Articles of Confederation until Virginia gave up its claims on western lands. Because of uncertain western borders, considerable jealousy existed between the states about how much land each state controlled—especially for new western states joining the Union.

Under the Articles of Confederation, land not devolved to the states was managed and essentially was assigned to private owners on a first-come, first-served system. The states bickered over land appropriations even after the Constitution was ratified. Boundary lines were determined by landmarks and often resulted in confusion and controversy.[3]

In order to generate badly needed revenue after the Revolutionary War, the U.S. government sold public land to private owners regularly. The primary purpose of those sales was to boost government revenue rather than to encourage settlement. The Land Ordinance of 1785 organized land sales, dividing the land into townships of thirty-six one-mile squares, or 640 acres. Sales were not very common; at $1 per acre for 640 acres, many couldn't afford the land. It took about twenty years (until 1800) for the government to resolve this problem. It reduced the minimum size of each saleable lot to 320 acres, increased the price per lot to $1.25 ($400 total) and allowed buyers to acquire it in four payments of $100 each. This made the land slightly more accessible, but not by much.[4]

In 1812, the General Land Office was formed under the U.S. Treasury Department to deal with the recording of public land sales. The office sold over 1 billion acres of federal public land to state governments and private landowners.[5] Because citizens were allowed to buy smaller parcels, more of them could afford to buy land, even though the per-acre price exceeded $1.25. During President James K. Polk's presidency, Treasury Secretary Robert J. Walker argued that the General Land Office really did not belong in the Treasury Department. Thus, the U.S. Department of the Interior was formed as a new home where the land office continued its business of overseeing public land sales.

President Andrew Jackson, a supporter of states' rights but also a promoter of the Union, vetoed a bill in 1833 that would grant lands and the proceeds of public land sales to "certain States" because the bill would appropriate the land only for a "limited time."[6] He asserted "the importance . . . of . . . a proper and final disposition of the whole subject of the public lands."[7] President Jackson's opinion reflected that of the public—that public lands should be transferred to states and private citizens. The land, however, was still too expensive for many citizens.

The Homestead Act of 1862 provided the opportunity to obtain title to public land without having to pay money up front. It allowed people to claim and settle 160 acres of land by living on it for five years, improving or cultivating it, and building a "12 by 14" structure on it. After five years, the papers proving ownership were filed, and a small fee was paid.[8] The law enabled more than 1.6 million homesteaders to claim land by 1934, further allowing the government to dispose of public lands to private owners.[9]

In total, about 10 percent of the country, or 270 million acres, were settled through homesteading.[10] The Homestead Act of 1862 allowed people to buy more land than they could have afforded at the higher prices. Land speculators took advantage of the low prices and a loophole in the law. The Homestead Act failed to specify the units of measurements of the required dwelling, so land speculators built tiny 12-by-14 inch structures to claim the land.[11]

The Push for Conservation

The notion that land and natural resources could be valuable merely by existing was unheard of prior to the Industrial Revolution and its increase in wealth and quality of life. As we explained in Chapter 4, the ideas of conservation and preservation entered public perception around the end of the nineteenth century, following the Industrial Revolution. One result is that public land policy flipped. The conservationists and preservationists claimed that transferring federal public land to private owners no longer was desirable; instead, public lands should remain in the hands of the federal government.

Two events marked this change in public opinion: The establishment of Yellowstone National Park in 1872, and the General Land Reform Act of 1891, which created the notion of "forest reserves."[12] Nicknamed the country's "Conservation President," Theodore Roosevelt spearheaded the fight for preservation of natural resources during his (one and a half) terms in office, declaring, "there can be no greater issue than that of conservation in this country."[13]

The U.S. Congress pushed for conservation again in 1934, with the passage of that year's Taylor Grazing Act. The law allotted 80 million acres of public land for grazing districts to be managed by the BLM and ended the era of free access by cattle and sheep ranchers.[14] The U.S. Grazing Service was established by that law, which eventually created thirty-four federal grazing districts in the western United States. The Taylor Grazing Act echoed the conventional wisdom at the time, holding that the federal government would not maintain ownership of public lands for eternity. The first sentence of the Taylor Grazing Act states that its purpose was to "promote the highest use of the public lands *pending its final disposal*" (emphasis added).[15] The federal government, then, did not intend to keep public land for itself indefinitely. Thirty years later, however, general opinions were changing, as illustrated by public lands historian Paul Wallace Gates:

> Many Americans take great pride in the national parks, enjoy the recreational facilities in the national forests, and in large numbers tour the giant dams and reservoirs of the Reclamation Service . . . National pride in the possession and enjoyment of these facilities seems to be displacing the earlier views.[16]

One piece of evidence that earlier views were being displaced is Woody Guthrie's folk song, "This Land is Your Land." Written in 1940 and first recorded in 1944, it became an alternative American anthem for many. The following verse is often left out but was included by Bruce Springsteen and Pete

Seeger, at We Are One, the Obama administration's inaugural celebration at the Lincoln Memorial on January 18, 2009.

There was a big high wall there that tried to stop me;
Sign was painted, it said private property;
But on the back side it didn't say nothing;
This land was made for you and me.

The song, written in response to Irving Berlin's "God Bless America," is a clear statement about who Guthrie thought ought to own America.[17] It also might be seen as corresponding closely to the views of President Obama as well as those of the conservationists and preservationists who became politically active at the time of John Muir and Teddy Roosevelt.

By the 1960s, Congress was divided between the newly environmentally aware politicians and those concerned with traditional public land uses like ranching, logging, and mining. In 1962 Congressman Wayne Aspinall, chairman of the House Interior and Insular Affairs Committee, suggested to President John F. Kennedy that a broad examination of public lands policy was overdue and asked the president for his input. President Kennedy agreed, and the result was legislation establishing the Public Land Law Review Commission in 1964 to examine public land management policies and to determine what changes, if any, were necessary. It took the commission six years to submit its report, entitled *One Third of the Nation's Land*. The commission was given the responsibility of judging whether public lands should "be (a) retained and managed or (b) disposed of, all in a manner to provide the maximum benefit for the general public."[18] After studying public land policy, the commission ultimately recommended an entirely new management plan, advising the repeal of hundreds of old land laws.

As a result of the report, the Federal Land Policy and Management Act (FLPMA) was passed in 1976. Its purpose, indicated on its title page is, "to establish public land policy; to establish guidelines for its administration; to provide for the management, protection, development, and enhancement of the public lands; and for other purposes."[19] The commission's report and its FLPMA sequel represented a new paradigm of federal land policy that resulted in substantial changes to federal land policy and remain the core of land policy nearly forty years later.

Federal Land Policy and Management Act

The FLPMA of 1976 was a monumental piece of legislation, as it superseded more than 3,000 federal land use laws, reversed the policy of transferring public lands to the states or private owners, and expanded access to land management decisions to a much wider public than traditional livestock and mining constituencies. After the new law, it was no longer correct to refer to the Bureau of Land Management (BLM) as the Bureau of Livestock and Mining, as its detractors had dubbed it. Although it is not as truly organic as the Forest Service's authorizing law, FLPMA fulfilled that role for the BLM.[20] FLPMA's multiple use mandates also apply to the U.S. Forest Service and the National Park Service.

The BLM manages more than 260 million acres of public land—nearly one-eighth of all of the United States. In addition, the BLM manages the 700 million acres of the federal subsurface mineral estate. FLPMA mandates that the BLM manage its portion of the federal estate according to four main standards: (1) maintaining federal land ownership, (2) multiple-use planning, (3) sustained yield, and (4) the unnecessary or undue degradation clause. The fourth mandate is defined in 1980 regulations as "surface disturbance greater than what would normally result when an activity is being accomplished by a prudent operator in usual, customary, and proficient operations of similar character and taking into consideration the effects of operations on other resources and land uses."[21]

These guidelines, particularly the ones concerning multiple public land uses, create a unique incentive pattern for people wishing to access or to determine federal land uses. Environmentalists want land to be preserved and protected, energy companies want to produce more energy, farmers want to cultivate the land, contractors want to build on the land, archaeologists want to preserve the cultural aspects of the land, and nearby towns want to use the land to attract tourists to boost their local economies. Because everybody wants something and resources are scarce, each group is incentivized to promote its parochial land management interests above those of others.

The BLM (or some other agency) is expected to impose fairness and order but instead adds to the already messy incentive structure. Agencies are far from neutral in the public land debate; they have their own interests and motives for public land management. The individuals who work in the agencies have vested interests in public lands and how they are used. If they didn't, they wouldn't be working for the agency. Thus, they are likely to have preferences that bleed into how they manage public land.

Federal Ownership of Public Land

Prior to 1976, the federal government consistently had sold, transferred, and traded tracts of public land. When the various states were admitted to the Union, the federal government promised, in each state's enabling act, to "extinguish title" or transfer the title of public lands to them "in a timely fashion."[22] FLPMA reneged on that promise by declaring that "public lands be retained in Federal ownership"[23] and repealing provisions of the Homestead Act of 1862. Declaring its intent to retain public lands in federal ownership marked a significant change from the promises made in states' enabling acts, despite the Supreme Court having called the enabling acts "trusts," "solemn compacts," and "bi-lateral agreements."[24]

Passing FLPMA didn't mean that public lands were to remain entirely public. The BLM asserts that "retention of lands does not exclude private interests from actually developing and using the resources on public lands."[25] The bureaucratic red tape, however, and the time required to process the paperwork to develop federal lands often result in much more conflict and effort than simply opting for private land ownership instead. Officials in Fruit Heights, Utah, learned this lesson through trial and error.

Fruit Heights Cemetery

While the BLM is the primary agency subject to FLPMA's requirements, the Forest Service also is required to comply with that law. The situation facing Fruit Heights, Utah, is a further illustration of FLPMA's problematic implementation.

Cemeteries in Utah, as elsewhere, are filling up. Before the 1830s, churches and other houses of worship provided land for graveyards on church property or town commons. As populations grew, however, the small church graveyards, already surrounded by houses and buildings, couldn't expand to meet the demand for burial space. As a solution, in the 1830s, the growing cities switched over to using spacious "park-style cemeteries."[26] But these larger cemeteries are filling up too, just like the church graveyards of the early 1830s.[27] Fruit Heights is one such town. Space in its cemetery is almost gone.

Fruit Heights has been turning to the nearby city of Kaysville to bury its dead, but that no longer is an option. Kaysville's population has grown, too, and city leaders are prohibiting Fruit Heights' citizens from interring their dead in its cemetery.[28] The city's burial options are shrinking as cities around it likewise are scrambling to find more cemetery space.

Fruit Heights is tucked against the mountains, land that is managed by the U.S. Forest Service. Under FLPMA, the land is federally owned and will remain so unless Congress authorizes a transfer. During the 113th Congress, Representative Rob Bishop proposed legislation, the Fruit Heights Lands Conveyance Act, to transfer to the city 100 acres of federal land to enlarge its cemetery. The bill passed the House of Representatives in 2013 and never passed the Senate.

Fruit Heights will likely have to find other ways to acquire land. The city might be able to exchange or buy land from the Forest Service. But, the city has no land to exchange, so Fruit Heights will have to find the money to buy it. Raising the tens of thousands of dollars needed to purchase 100 acres will be formidable for a community with a population of only 5,000 people.

As illustrated by this case, it isn't easy to use or acquire public land in the hands of the federal government; in this case, it takes an act of Congress. The FLPMA requires the BLM and other federal agencies to manage public lands in the nation's best interest, but what tools do they actually have to gauge the nation's best interest? Fruit Heights' request is still in the preliminary stage. Even if the town raises enough money to buy the land it needs, the Forest Service and BLM must still decide whether the transfer is in the national interest. The BLM and Forest Service have full discretion under FLPMA and other acts (e.g., General Land Exchange Act of 1922, Townsite Act of 1958, and Small Tracts Act of 1938) to determine whether the public land will be more beneficial in Fruit Heights' possession.[29] Fruit Heights is entirely at the mercy of the agencies' decisions.

In managing public land, casualties often result from varying and often strict interpretation of the guidelines. The case of Gerald Chaffin, described at the beginning of this appendix, was one of these casualties. Ironically, just four years after he torched his house, the Small Tracts Act was passed, ensuring that if someone encroaches unknowingly on public land, the Forest Service can transfer the land to the private owner.[30]

That incident and others like it led to widespread resentment among many supporters of the 1960s' Sagebrush Rebellion, who protested against federal control of public lands within individual states. Nevadans were especially vocal, insisting that federal policy usually didn't benefit western states, where the majority of federal lands exist. Eighty-seven percent of Nevada, for example, is federally owned land.[31] In 2012, Utah legislators reignited the Sagebrush Rebellion, presenting House Bill 148, which would transfer control of federal lands to the state.[32] House Bill 148 outlines the legal reasoning of the proposal and the process through which state control would occur.

Multiple Use

The FLPMA gives the BLM and other agencies general guidelines for managing public lands. One of the most prominent is established in Section 102(a) of the law, proclaiming that land management be grounded in "multiple use." Borrowing from a similar provision in the Multiple Use and Sustainable Yield Act of 1960, Congress defined *multiple use* as managing public lands and their resources in the best possible ways, benefitting the American people's current and potential future needs. The Federal Land Policy and Management Act essentially assigns the BLM as mediator over all the possible interdisciplinary approaches to the public land it manages, including "physical, biological, economic, and other sciences."[33] The law's Declaration of Policy identifies a list of factors that are considered legitimate and protected under multiple use:

> scientific, scenic, historical, ecological, environmental, air and atmospheric, water resource, and archeological values . . . preserv[ing] and protect[ing] certain public lands in their natural condition . . . provid[ing] food and habitat for fish and wildlife and domestic animals; and . . . provid[ing] for outdoor recreation and human occupancy.[34]

Protection and preservation are emphasized in this list, and a subsequent clause stresses environmental protection even more. The clause stipulates that public land deemed of "critical environmental concern" should be regulated and planned for quickly.[35] These lands are called Areas of Critical Environmental Concern.

This assertion of environmental and scientific protection and uses is followed by a condition also to provide for the "Nation's need for domestic sources of minerals, food, timber, and fiber from the public lands including implementation of the Mining and Minerals Policy Act of 1970 . . . as it pertains to the public land."[36] While preservation is important, FLPMA stipulates that public land should also be used for more than just preservation, as stated in the mandate of sustained yield.

On public lands, one response to the problem of allowing variety has been to mandate multiple uses. Managing for multiple uses, however, does not necessarily mean better management of the land. Instead, it means that all lands are managed for all uses. In one public meeting we attended, Marion Clawson, the first director of the BLM, argued that the result of multiple use management has been, "a little of everything everywhere, without regard to cost or effect." Furthermore, the management process has given particular organized interests a means to enforce their views on others. For example, motorized recreationists champion multiple use to justify their own preferences at the expense of those seeking solitude. Similarly, logging companies justify timber harvesting on uneconomical forestlands as being mandated by multiple use policy. They claim that logging roads, after all, benefit recreationists and wildlife.

If political and structural problems make agencies inflexible about allowing new data or analytical methods to intrude on their management practices, inflexibility toward introducing new uses of a resource is even more severe. Although people's preferences change with time, such as the huge increase in the number of people who view the national forests as amenity sheds rather than timber sheds, agencies are extremely slow to follow. Administrators wish to avoid controversy, and introducing competing uses for the land can trigger political conflict. Those groups wielding the most political power, the best organization, and the most influence can rule in such a situation, regardless of the effects.

The Southern Utah Wilderness Alliance (SUWA) is a collection of political entrepreneurs who have been successful at pushing back against multiple use management practices. They rely heavily on litigation (and its threat) to achieve their goals. SUWA's purpose is "to defend America's red-rock wilderness from oil and gas development, unnecessary road construction, rampant off-road vehicle use, and other threats to Utah's wilderness-quality lands."[37] Wilderness-quality lands are managed as if they were wilderness, so SUWA's goal is non-use, as opposed to multiple use. Non-use may be too strong a term because some uses are allowed in wilderness, but uses are greatly restricted. The purpose of preventing other uses is rooted in the balance of nature ideology, attempting to freeze nature in its current state or return it to some arbitrary point that SUWA has decided is nature's baseline.[38]

In 1999, SUWA filed suit against the BLM (*Southern Utah Wilderness Alliance v. Norton*), alleging that the agency had "violated the Federal land Policy and Management Act . . . and the National Environmental Policy Act . . . by not properly managing off-road vehicle and/or off-highway vehicle . . . use on federal lands."[39] Under FLPMA, the BLM had designated 2.5 million acres of land in Utah as wilderness study areas, which required the agency to manage the land "so as not to impair the suitability of such areas for preservation as wilderness."[40] SUWA asserted that the BLM failed to assess and even ignored off-road vehicles when issuing use permits for the wilderness study areas. The case was later settled in the U.S. Supreme Court; Justice Antonin Scalia noted that it wasn't the court's job to micromanage the BLM's management of public lands.[41] *SUWA v. Norton* is significant because it reestablishes the BLM's authority in interpreting FLPMA and how best to manage public land.

As we noted in Chapter 2, political entrepreneurs are not always successful, as in the case of *SUWA v. Norton*. Successful private entrepreneurs often are serial entrepreneurs, succeeding sometimes and failing other times. SUWA is an example of a successful public serial entrepreneur. The coalition lost in *Norton* but prevailed in other cases. In 2014, SUWA, along with attorneys from Earthjustice, won a suit against the National Park Service that had been filed originally in 1995. SUWA's description of the issue follows:

> In 1995, SUWA attorneys (including now-SUWA Executive Director Scott Groene) filed a lawsuit challenging a decision by the National Park Service to permit jeep use in Salt Creek Canyon despite overwhelming evidence that such use was destroying cultural sites and the

fragile stream. In the nearly 20 years that followed the filing of that lawsuit SUWA's legal team and dogged members never gave up the fight to protect this special place. Now we can reap the fruits of that labor: the knowledge that Salt Creek Canyon will be preserved for current and future generations as a refuge of quiet beauty.

As our former colleague and now partner at Earthjustice Heidi McIntosh said, SUWA is about endless pressure applied endlessly. The Salt Creek decision is a testament to that principle.[42]

Salt Creek Road is a twelve-mile road that is the primary way for tourists to reach several highly prized destinations in Canyonlands National Park, the most notable of which is Angel Arch, nine miles from where the road begins. The road crosses the streambed multiple times and sometimes is in the streambed. By succeeding in getting the road closed, SUWA has made Angel Arch accessible only to those who can walk the nine miles. We are not judging whether or not the road should have been closed to vehicles. We are just noting that the court's decision was the result of long-term, successful political entrepreneurship.

According to SUWA's website, one of its causes is to push President Obama to declare the Greater Canyonlands in southeastern Utah a national monument. This act would "protect the 1.4 million acres of public BLM land surrounding Canyonlands National Park with the stroke of a pen"[43] Doing so would increase regulations and protections on the land, "[stopping] any uranium mining and oil and gas development."[44] The group's website invokes Muir's emotional, even religious zeal:

> There are few enough places left in the lower 48 where we can truly lose ourselves, stand alone and bask in creation's splendor. One of them is the Greater Canyonlands region, a stretch of matchless country in southern Utah at the heart of which is Canyonlands National Park.[45]

Multiple use groups have organized to counter SUWA and although they invoke FLPMA's multiple use mandate, they are at a strategic disadvantage because it is hard to argue that God rides an ATV or that cows grazing in "a stretch of matchless country" or worse yet, oil and gas exploration and development, do not detract from "creation's splendor."

Sustained Yield

The sustained-yield mandate is premised on the idea that land should be used to benefit the national interest.[46] The term *sustained yield* suggests the importance of obtaining resources from the land continually. Having and protecting land and resources is a potentially valid approach, but pushing production forever into the future is not consistent with a sustained yield approach. This, by the way, is a modern visitation of the debate between John Muir and Gifford Pinchot described in Chapter 2.

The idea of a single appropriate sustained yield is steeped in balance of nature rhetoric. Nature, according to the theory, is drifting around a baseline. Thus, it is assumed that when a species or resource is in abundance, a set amount can recurrently be harvested or extracted without hurting the foundational baseline. This idea was first established in the commercial fishing industry.

Around the 1930s, scientists started to observe significant effects of open access fishing of fish populations. They began to worry about what we would today call a tragedy of the commons. E. S. Russell, Britain's director of fisheries investigations, was one of the first to claim that fishing may be reducing the numbers of fish in the ocean.[47] W. F. Thompson, Director of the International Fisheries Commission, published similar findings, illustrating a decline of fish populations from fishing. Russell's and Thompson's studies and their evidence of declining fish inspired British scientist Michael Graham to assert, "Fisheries that are unlimited become unprofitable."[48] In other words, a profitable fishery needed to limit the number of fish taken from the ocean. Those limits would actually benefit fisherman. The novel idea developed and "maximum sustained (or sustainable) yield" was the result.

The 1940s and 1950s were replete with praise for the "scientifically sound" idea of maximum sustained yield, which would save our fisheries and manage them perfectly forever. Maximum sustained yield was determined for each species of fish, ensuring that the highest possible rate of sustained fishing a population could withstand was allowed—but no more. Author and ecologist Daniel Botkin explains:

> Maximum Sustainable Yield is a concept that comes out of the logistic growth equation, which says that a population, undisturbed by people,

will grow to a fixed carrying capacity and its abundance will stay at that carrying capacity forever. A "logistic" population reaches its maximum growth rate when its population is exactly at one half of it carrying capacity. The logistic has no environmental variables. As applied in standard fisheries management models, the only variable is the amount of human harvest. The logistic curve is therefore a mathematically explicit statement of the ancient balance of nature idea, which goes back to the ancient Greek philosophers, that nature undisturbed achieves a constancy.[49]

As long as fishing was kept at or below the maximum sustainable line, the species was expected to continue thriving forever. If a perfect balance exists around which fish populations hover without interruption, then we can fish down to that balance or right above it, and the fish will continue reproducing at that line. The theory was lauded as a scientifically accurate solution to the problem of overfishing. The problem is that Russell, Thompson, and their peers were relying on data from only one time period. In the scope of the total human impact on fish, they had barely any knowledge at all as to how fishing really affected fish populations.

In 1996, six West Coast fisheries collapsed as a result of overfishing. The U.S. secretary of commerce called it a federal disaster.[50] Some turned to the tragedy of the commons as the cause of the collapse, but how could that be? Fishermen had been adhering to the maximum sustained yield for forty years, and they had done so *before* the population would show any signs of being overfished—or so they thought.

The fundamental problem with this theory is the same problem found in the balance of nature ideology: nature changes constantly. There is no stationary line around which nature hovers. Botkin continues:

> no data about any aspect of the environment follows the logistic curve. Nature in all its characteristics—climate, oceans, ecosystems, species, populations—is always changing. There are well-known mathematical methods that can deal with such varying systems, and they should be applied instead of the maximum sustainable yield and its mathematical basis.[51]

Since nature is constantly changing, using outdated methods to manage nature that assume stability can produce good results only serendipitously. Pulling fish out of the ocean at a rate set in 1950 does not necessarily ensure that the fish population will remain at some fixed equilibrium.

If an agency relies too heavily on the idea that public land should be managed to return to its state at a specific point in time, the results can be disastrous. The BLM has had some experience with this possibility: It manages the wild horses roaming western public land. The Wild Free-Roaming Horses and Burros Act requires the agency to "protect and manage wild free-roaming horses and burros as components of the public lands . . . in a manner that is designed to achieve and maintain a thriving natural ecological balance on the public lands."[52] The agency relies heavily on multiple use and sustained yield in deciding if horses or burros should be removed and, if so, how many. In addition, the BLM is required to control for overpopulation on public lands and "immediately remove excess animals from the range so as to achieve appropriate management levels."[53] The BLM has determined these appropriate management levels much like a maximum sustainable yield is determined; it is the "estimated number of wild horses the land can support while maintaining a thriving natural ecological balance with other resources and uses."[54]

In 2009 the Calico Mountains Complex, a compilation of five herd management areas for wild horses in Nevada, exceeded the appropriate management level by more than 2,000 horses.[55] In compliance with FLPMA and the Wild Horses Act, the BLM proceeded to gather wild horses and house them in facilities in South Dakota, Kansas, and Oklahoma. Interior Secretary Ken Salazar was sued for the action. The suit argued that numerous laws, including the Wild Horses Act and the National Environmental Policy Act, were violated in the BLM's removal of the wild horses.[56] The case was decided on the basis of technicalities, and the BLM was allowed to continue its activities.

By using this system, the BLM is setting up uninformed, albeit carefully calculated, guidelines that may affect the wild horse population adversely. Continuing to view nature as static and clinging to an unchanging quota is risky, because the BLM doesn't actually know at what level wild horse populations should be managed to keep them healthy and thriving. Despite all this, the BLM is given full discretion in managing land, especially in the unnecessary or undue degradation clause of FLPMA.

Title III of FLPMA details the BLM's responsibilities and guidelines for land management. The law requires the BLM to employ multiple-use and sustained-yield practices to manage public land, but the agency must also prevent all "unnecessary or undue degradation."[57] If the BLM is adhering to the multiple-use and sustained-yield provisions of FLPMA, then they'll most likely keep "unnecessary or undue degradation" from taking place.[58] This phrase, including the multiple-use and sustained-yield mandates, gives the BLM and other agencies considerable freedom as to how to manage public lands; FLPMA gives the agency authority to interpret "multiple use," "sustained yield," and "unnecessary or undue degradation" as they see fit.

As noted previously, agency discretion in matters such as these won't always align with "the nation's best interest," local interests, or the interests of those who actually use agency lands. There is simply no mechanism with which to gauge whether or not the agency is meeting the mandate to "avoid unnecessary and undue degradation" or acting in the nation's best interest.

Conclusions

Public land managers are locked in the past by mandates requiring management according to an arbitrary historical baseline. They are also caught in the balance of nature ideology that generated many of the laws they must enforce and carry out. Innovation and experimentation in management are unlikely, given those mandates and the guiding ideology. Besides, such actions will be fought in Congress, the courts, and the press. The safe thing for managers to do, in terms of their own careers, but not necessarily in terms of environmental betterment, is to stay safely within the balance of nature and preservation ideologies that are embedded in public land law.

Notes

Chapter 1

1. Charles T. Rubin, *The Green Crusade* (New York: The Free Press, 1994).
2. Basgall, M. (1996). Defining a New Ecology. *Duke Magazine*, 39–41. Retrieved May 29, 2015.
3. Audrey Hudson, "Environmentalists Oppose Obama Plan to Develop Solar Energy," *Human Events* (September 13, 2012).
4. Outhwaite, W. (Ed.). (n.d.). The Blackwell Dictionary of Modern Social Thought (p. 201). Malden, Massachusetts.
5. Simmons, R., Yonk, R., & Thomas, D. (2011). Bootleggers, Baptists, and Political Entrepreneurs; Key Players in the Rational Game and Morality Play of Regulatory Politics. The Independent Review, 15(3), 367–381. Retrieved May 28, 2015, from https://www.independent.org/pdf/tir/tir_15_03_3_simmons.pdf
6. Kingdon, J. (1984). Agendas, Alternatives, and Public Policies. New Jersey: Addison-Wesley Educational.
7. Dahl, R. (1956). A Preface to Democratic Theory (p. 146). Chicago, Illinois: University of Chicago Press.
8. Peter Kareiva, Michelle Marvier, and Robert Lalasz, *Conservation in the Anthropocene: Beyond Solitude and Fragility* (2012).

Chapter 2

1. Daniel B. Botkin, 2012. *The Moon in the Nautilus Shell: Discordant Harmonies Reconsidered.* (New York: Oxford University Press, 2012), xii.
2. The balance of nature is a belief system, an assumption, and an ideology, depending on the context of its use. When used in policy and political contexts, we will refer to it as an ideology.
3. Corinne Zimmerman and Kim Cuddington, 2007, "Ambiguous, Circular and Polysemous: Students' Definitions of the 'Balance of Nature' Metaphor," *Public Understanding of Science.* 16 (October 2007): 393–406; quotation from page 393.
4. William Cronon, *Uncommon Ground: Toward Reinventing Nature.* (New York: W. W. Norton and Company, 1995), 24.
5. Barbour, M. (1995). Ecological Fragmentation in the Fifties. In W. Cronan (Ed.), Uncommon Ground. New York: W.W. Norton & Company.
6. World Wildlife Fund, "Ecological Balance," WWF Global website.
7. Rachel Carson, 1962. *Silent Spring.* (Houghton Mifflin, 1962), 246, 5.
8. Ibid., 5.
9. Barry Commoner, *The Closing Circle: Nature, Man, and Technology* (New York: Knopf, 1971); Donella H.

Meadows, Club of Rome, Potomac Associates, *The Limits to Growth* (Universe Books, 1972); Albert Gore, *Earth in the Balance: Ecology and the Human Spirit* (Rodale, 2006); and James E. Lovelock, *The Ages of Gaia: A Biography of Our Living Earth* (New York: Norton, 1988); *Gaia: The Practical Science of Planetary Medicine* (Oxford University Press, 1991), *Healing Gaia: Practical Medicine for the Planet* (Harmony Books, 1991); and *The Revenge of Gaia: Why the Earth is Fighting Back and How We Can Still Save Humanity* (Penguin Books Limited, 2007).

10. John C. Kricher, *The Balance of Nature: Ecology's Enduring Myth* (Princeton University Press, 2009).

11. Evelyn C. Pielou, *After the Ice Age: The Return of Life to Glaciated North America* (Chicago: University of Chicago Press, 1991).

12. Paul Schullery, *Searching for Yellowstone: Ecology and Wonder in the Last Wilderness* (Helena, MT: Montana Historical Society Press, 2004), 18.

13. What happened to the buffalo. (2009, March 5). Retrieved May 27, 2015, from http://www.americanbison.org/

14. Lynn Jacobs, *Waste of the West: Public Lands Ranching* (Tucson, AZ: Lynn Jacobs, 1991), 118; Ray Rasker, Norma Tirrell, and Deanne Klopfer. 1991. *The Wealth of Nature: New Economic Realities in the Yellowstone Region* (Washington, DC: Wilderness Society, 1992), 63; Charles E. Kay, "Aboriginal Overkill and the Biography of Moose in Western North America," *Alces*. 33 (1997): 141–164.

15. Peter S. Ogden, "Peter Skene Ogden's Snake Country Journals," 1824–25 and 1825–26, in *Hudson's Bay Record Society Publication, 8* (1950), ed. E. E. Rich and A. M. Johnson, 73.

16. Charles E. Kay, Yellowstone's Northern Elk Herd: A Critical Evaluation of the "Natural Regulation" Paradigm (Ph.D. Dissertation, Utah State University, Logan, Utah, 1990); *Wolf Recovery, Political Ecology, and Endangered Species* (Oakland, CA: Independent Institute, 1996); "Viewpoint: Ungulate Herbivory, Willows, and Political Ecology in Yellowstone," *Journal of Range Management,* 50 (1997): 139–145; "Yellowstone: Ecological Malpractice," *PERC Reports* 15(2) (1997).

17. Charles E. Kay and C. W. White, "Long-Term Ecosystem States and Processes in the Central Canadian Rockies: A New Perspective on Ecological Integrity and Ecosystem Management," in R. M. Linn (Ed.), *Sustainable Society and Protected Areas,* ed. R. M. Linn (Hancock, MI: The George Wright Society, 1995), 119–132; Charles E. Kay, B. Patton, and C. White, *Assessment of Long-Term Terrestrial Ecosystem States and Processes in Banff National Park and the Central Canadian Rockies* (Banff, AB: Resource Conservation, Parks Canada, Banff National Park, 1994).

18. Charles E. Kay, 1996(a). "Ecosystems Then and Now: A Historical Approach to Ecosystem Management (Occasional Paper No. 23)," in *Proceedings of the Fourth Prairie Museum of Alberta Natural History,"* ed. W. D. Wilms and J. F. Dormarr, 87–89.

19. Kay, "Yellowstone's Northern Elk."

20. R. Daubenmire, "The Western Limits of the Range of American Bison," *Ecology* 66 (1985): 622–624; Donald K. Grayson, *The Desert's Past: A Natural Prehistory of the Great Basin* (Washington, DC: Smithsonian Institute Press, 1993).

21. Stearn, Esther W., Allen E. Stearn. 1945. *The Effect of Smallpox on the Destiny of the Amerindian*. Boston, MA: Bruce Humphries; Noble D. Cook and W. G. Lovell, eds., *Secret Judgments of God: Old World Disease in Colonial Spanish America* (Norman: University of Oklahoma Press, 1992).
22. Henry F. Dobyns, *Their Numbers Become Thinned: Native American Population Dynamics in Eastern North America* (Knoxville: University of Tennessee Press, 1983).
23. Ann F. Ramenofsky, *Vectors of Death: The Archaeology of European Contact* (Albuquerque: University of New Mexico Press, 1987).
24. David E. Stannard, *American Holocaust: Columbus and the Conquest of the New World* (New York: Oxford University Press, 1992).
25. Marvin T. Smith, *Archaeology of Aboriginal Culture Change in the Interior Southeast* (Gainesville: Florida State Museum, 1987).
26. Jared Diamond, "The Golden Age that Never Was," *Discover* 9(12) (1988): 70–79.
27. Charles C. Mann, *1491: New Revelations of the America's Before Columbus* (New York: Vintage Books, 2006).
28. Ibid., 366.
29. Ibid., 366. For an extended discussion of this and similar issues see *Wilderness and Political Ecology: Aboriginal Influences and the Original State of Nature*, C.E. Kay and R.T Simmons, eds. University of Utah Press. 2002.
30. Valerius Geist, *Buffalo Nation: History and Legend of the North American Bison* (McGregor, MN: Voyageur Press, 1996).
31. Hickerson, H. The Virginia Deer and Intertribal Buffer Zones in the Upper Mississippi Valley. In A. Leeds & A. Vayda (Eds.), *Man, Culture, and Animals: The Role of Animals in Human Ecological Adjustment*. American Association for the Advancement of Science. (1995): 45.
32. Charles E. Kay, "Aboriginal Overkill: The Role of Native Americans in Structuring Western Ecosystems," *Human Nature* 5 (1994): 359–398; "Aboriginal Arguments." *Journal of Forestry*. 95 (1997): 8.
33. Meriwether Lewis, William Clark, and Thomas Jefferson. *History of the Expedition Under the Command of Lewis and Clark: To the Sources of the Missouri River, Thence Across the Rocky Mountains and Down the Columbia River to the Pacific Ocean, Performed During the Years 1804–5–6*, Volume 3 (Henry Stevens & Son, 1893), 1197.
34. F. W. Raynolds, *Report on the Exploration of the Yellowstone River in 1859–60*, Senate Executive Document 77, 40th Congress, Second Session, 1868.
35. D. Martinez, "Back to the Future: Ecological Restoration, the Historical Forest, and Traditional Indian Stewardship" (paper presented at a Watershed Perspective and Native Plants Conference, Olympia Washington, February 26, 1993).
36. Jonathan S. Adams and Thomas O. McShane, *The Myth of Wild Africa: Conservation Without Illusion* (New York: W. W. Norton, 1992).
37. Yellowstone's Northern Range; Where Nature Takes Its Course. (n.d.). Retrieved May 27, 2015, from http://www.nps.gov/yell/planyourvisit/upload/paper.pdf
38. See Mann, *1491*.
39. R. E. Grumbine, "What is Ecosystem Management?" *Conservation Biology* 8(1) (March 1994): 27–34; Stephen Budiansky, *Nature's Keepers: The New*

Science of Nature Management (New York: Free Press, 1995).
40. Budiansky, *Nature's Keepers.*
41. Tim Caro, *Conservation by Proxy: Indicator, Umbrella, Keystone, Flagship, and Other Surrogate Species* (Washington, DC: Island Press, 2010).
42. Botkin, *The Moon in the Nautilus Shell.*
43. Pielou, *After the Ice Age,* 101. Andy Kerr, former Executive Director of the Oregon Natural Resources Defense Council, suggested in a 1998 conference we attended that "ecosystem" is in fact a marketing term, not a scientific term.
44. Emma Marris, *Rambunctious Garden: Saving Nature in a Post-Wild World* (New York: Bloomsbury USA, 2011).
45. Botkin, *The Moon in the Nautilus Shell,* 70.
46. Ibid.
47. Ibid.
48. Mann, *1491;* Botkin, *The Moon in the Nautilus Shell*; Budiansky, *Nature's Keepers.*
49. U.S. Forest Service. United States Department of Agriculture, *Landmark Restoration Effort to Heal Northern Arizona's Forests and Communities,* retrieved in 2013 from: https://fs.usda.gov/Internet/FSE_DOCUMENTS/stelprdb5158926.pdf
50. Thomas M. Bonnicksen, "An Analysis of a Plan to Maintain Old-Growth Forest Ecosystem" (American Forest and Paper Association, *Forest Resources Technical Bulletin,* TB 94-3, 1994).
51. Peter Dominic Adrian Teensma, John T. Rienstra, and Mark A. Yeiter, "Preliminary Reconstruction and Analysis of Change in Forest Stand Age Classes of the Oregon Coast Range From 1850 to 1940," *USDAI Bureau of Land Management Technical Note* T/N OR-9 (1991); Robert Zybach, "Oregon's 2012 Wildfires: Predictable and Preventable," *Oregon Fish & Wildlife Journal,* 34 (4) (2012): 7–17. Retrieved from: http://www.nwmapsco.com/ZybachB/Articles/Oregon_Wildfires_2012/Zybach_2012c.pdf
52. Robert Boyd, "Strategies of Indian burning in the Willamette Valley," *Canadian Journal of Anthropology* 5 (1986): 65–86.
53. D. MacCleery, "Understanding the Role That Humans Have Played in Shaping America's Forest and Grassland Landscapes," *Forest Service Department of Agriculture Newsletter,* September 7, 1994.
54. Thomas M. Bonnicksen, "An Analysis of a Plan to Maintain Old-Growth Forest Ecosystem," American Forest and Paper Association, *Forest Resources Technical Bulletin* TB 94-3 (1994).
55. In most western forests, frequent low-intensity ground fires were once the norm while stand-replacing crown fires generally occurred only at higher elevations and in wetter environments, such as the Pacific Northwest (Bonnicksen, 1993). With fire suppression and fire exclusion, however, forest fuels have both grown up and accumulated to unprecedented levels. Thus, crown fires are now common where they never occurred in the past, while in other areas, the size and intensity of the crown fires has increased since earlier times. Even in wet coastal environments, though, aboriginal burning was still once widespread (Norton, 1979; Boyd, 1986; Turner, 1991; Gottesfeld, 1994).
56. Marris, E. (n.d.). Emma Marris. Retrieved May 29, 2015, from http://emmamarris.com/?page_id=17

Chapter 3

1. John E. Mueller, *Capitalism, Democracy, and Ralph's Pretty Good Grocery* (Princeton, NJ: Princeton University Press, 1999) 137.
2. See Chapter 8 on the Endangered Species Act for more information.
3. Quoted in David Blackmon, "The 'Sue and Settle' Racket," *Forbes,* May 27, 2013.
4. Mark Dowie, *Losing Ground: American Environmentalism at the Close of the Twentieth Century.* (Cambridge: The MIT Press, 1996), 1.
5. Peninah Neimark and Peter Rhoades Mott, *The Environmental Debate: A Documentary History, with Timeline, Glossary, and Appendices* (2nd ed.), (Amenia, NY: Grey House, 2011), 190.
6. Andy Reynolds, *A Brief History of Environmentalism*; Joseph E. Taylor III and Matthew Klingle, "Environmentalism's Elitist Tinge has Roots in the Movement's History," *Grist,* March 9, 2006.
7. "Cuyahoga River Fire," *Ohio History Central*, n.d.
8. Ibid.
9. Miles Corwin, "The Oil Spill Heard 'Round the Country!" *Los Angeles Times,* January 28, 1989.
10. Gilbert Cruz, "Top 10 Environmental Disasters," *Time,* May 3, 2010.
11. Quoted in Corwin, "The Oil Spill."
12. Jonathan Adler, "The Fable of Federal Environmental Regulation: Reconsidering the Federal Role in Environmental Protection," *Case Western Reserve Law Review,* 55 (September 3, 2004): 96–101.
13. Rachel Carson, *Silent Spring* (reprint ed.), Boston: Houghton Mifflin, 2002), 6
14. Ibid., 15.
15. Quoted in Robert Stone, *Earth Days* (film) (U.S.: American Experience, 2009).
16. J. E. de Steiguer, *The Origins of Modern Environmental Thought.* Tucson: University of Arizona Press, 2006, 40–41.
17. "Paul Müller–Biographical," The Nobel Foundation.
18. G. Fischer, Award Ceremony Speech, 1948, The Nobel Foundation.
19. Henry I. Miller and Gregory Conko, "Rachel Carson's Deadly Fantasies." *Forbes,* September 5, 2012.
20. Ibid.
21. John Berlau, *Eco-Freaks: Environmentalism is Hazardous to Your Health* (Nashville, TN: Nelson Current, 2006), 56.
22. Ibid., 57
23. Quoted in Berlau, *Eco-Freaks,* 39.
24. Paul Ehrlich, *The Population Bomb* (New York: Ballantine Books, 1968).
25. Quoted in Stone, *Earth Days.*
26. Ehrlich, *Population Bomb,* 15.
27. Taylor and Klingle, "Environmentalism's Elitist Tinge."
28. De Steiguer, *Origins,* 111.
29. Quoted in Stone, *Earth Days.*
30. Taylor and Klingle, "Environmentalism's Elitist Tinge."
31. Stewart L. Udall, The Quiet Crisis (New York: Holt, Rinehart and Winston, 1963).
32. De Steiguer, *Origins,* 54.
33. Bill Moyers, "Stewart Udall Biography," *NOW with Bill Moyers. Science and Health: Preserving the Parks* (PBS, November 14, 2003). Retrieved from: http://www.pbs.org/now/science/udall.html
34. Adler, "The Fable of Federal Environmental Regulation."
35. Ibid., 95.

36. Ibid., 99. See Chapter 8 on the Clean Water Act for more details on the Cuyahoga River Fire.
37. Ibid., 101.
38. Ibid., 96.
39. Michael Rotman, "Cuyahoga River Fire," *Cleveland Historical*.
40. Adler, "Fable," 96–97.
41. Ibid., 98.
42. David B. Johnson, 1991. *Public Choice: An Introduction to the New Political Economy* (Mayfield Publishing Co, 1991), 285.
43. Associated Press. "Report: Elk Not Overgrazing," *Missoulian*, March 13, 1995, C2.
44. Richard Behan, "RPA/NFMA—Time to Punt," *Journal of Forestry* (December 1981): 802–805.
45. R. B. Keigley, "An Increase in Herbivory of Cottonwood in Yellowstone National Park," *Northwest Science* 71 (1997): 127–136; and Statement of Richard Keigley to the Subcommittee on National Parks and Public Lands. Oversight Hearing on Science and Resource Management in the National Park System.
46. Clifford, F. (1993, November 22). Scientists Fight Over Who's Faithful to Yellowstone: Wildlife: Biologists charge that their findings have been suppressed. Park Service officials say they tolerate dissent. *Los Angeles Times*. Retrieved May 28, 2015, from http://articles.latimes.com/1993-11-22/news/mn-59658_1_national-park-service
47. Keigley, Statement. Simmons attended the subcommittee hearing and was told by the committee's staff director that the National Park Service had refused the subcommittee's request for Keigley to testify. Thus, the committee had to use their subpoena power in order to hear Keigley.
48. Alston Chase, "The Dark Side," *Range* 5(2) (1997): 4–8, 62.
49. Keigley, "An Increase in Herbivory."

Chapter 4

1. William Cronon, *Uncommon Ground: Rethinking the Human Place in Nature* (New York: W. W. Norton, 1996), 70.
2. Marris, *Rambunctious Garden*, 18.
3. Roderick F. Nash, *Wilderness and the American Mind* (4th ed.). (New Haven, CT: Yale University Press, 2001), 24
4. Neimark & Mott, *The Environmental Debate*, 70.
5. Cronon, *Uncommon Ground* (1996), 76. The term *wilderness cult* comes from Roderick Nash's book *Wilderness and the American Mind* (New Haven, CT: Yale University Press, 1967).
6. Cronon, *Uncommon Ground* (1996), 75.
7. Marris, *Rambunctious Garden*, 21.
8. Frederick Jackson Turner, "The Significance of the Frontier in American History," *American Historical Association. Chicago Worlds Fair.* Chicago, July 12, 1893. E 179.5.T958 1966.
9. Marris, *Rambunctious Garden*, 22.
10. Carl Russell, "Coordination of Conservation Programs," *Regional Review* 6 (1–2) (1941): 13–19.
11. Theodore Roosevelt Association, "Brief Biogrphy," 2013.
12. Nash, *Wilderness and the American Mind*, 4th ed., 125.
13. Taylor and Klingle, "Environmentalism's Elitist Tinge."
14. Mark Dowie, "Conservation: Indigenous People's Enemy No. 1?" *Mother Jones*, November 25, 2009.
15. Nash, *Wilderness and the American Mind*, 4th ed., 129.
16. Cronon, *Uncommon Ground*, 1996, 79.

17. Taylor and Klingle, "Environmentalism's Elitist Tinge."
18. William T. Hornaday, *Our Vanishing Wildlife* (New York: New York Zoological Society, 1913).
19. Taylor and Klingle, "Environmentalism's Elitist Tinge."
20. Ibid.
21. Robert H. Nelson, *Public Lands and Private Rights: The Failure of Scientific Management* (Lanham, MD: Rowman & Littlefield, 1995), 48, as quoted in Berlau, *Eco-Freaks*, 160.
22. Nash, *Wilderness and the American Mind*, 4th ed., 161.
23. Ibid., 162.
24. Cronon: *Uncommon Ground*, 1996, 72.
25. De Steiguer, *The Origins*, 12–13.
26. Nash, *Wilderness and the American Mind*, 4th ed., 161.
27. Ibid.
28. Ibid.
29. Ibid.
30. Hetch Hetchy: timeline of the ongoing battle over hetch hetchy, Retrieved June 23, 2015 from http://vault.sierraclub.org/ca/hetchhetchy/timeline.asp
31. Nash, *Wilderness and the American Mind*, 4th ed., 167.
32. Cronon, *Uncommon Ground*, 1996, 72.
33. Aldo Leopold, "Wilderness as a Form of Land Use," *Journal of Land and Public Utility Economics* 1(4) (1925): 398–404.
34. James M. Turner, *The Promise of Wilderness: American Environmental Politics Since 1964* (Seattle: University of Washington, 2012), 24–25.
35. De Steiguer, *The Origins*, 14.
36. "The Leopold Legacy: A Sand County Almanac." The Aldo Leopold Foundation, n.d. Retrieved from: http://www.aldoleopold.org/AldoLeopold/almanac.shtml; and Dowie, *Losing Ground*, 18.
37. As quoted in Dowie, *Losing Ground*, 19.
38. Ibid.
39. Stone, *Earth Day*.
40. Ibid.
41. Michael McCarthy, Michael. 2012, June 12. "Earthrise: The Image That Changed Our View of the Planet," *The Independent*, June 12, 2012.
42. Turner, *The Promise of Wilderness*. 97.
43. Ibid.
44. Stone, *Earth Days*.
45. Brad Knickerbocker, "Environmental 'Magna Carta' Law Under Fire," *Christian Science Monitor*, November 7, 2002. Retrieved from: http://www.csmonitor.com/2002/1107/p02s02-usgn.html
46. Turner, *The Promise of Wilderness*, 105.
47. Environmental Protection Agency, "The Guardian: Origins of the EPA," 1992. Retrieved from: http://www2.epa.gov/aboutepa/guardian-origins-epa
48. Stone, *Earth Days*.
49. Richard Nixon, "Statement About the National Environmental Policy Act of 1969, January 1, 1970." Retrieved from: http://www.presidency.ucsb.edu/ws/?pid=2557.
50. Dowie, *Losing Ground*, 33.
51. Turner, *The Promise of Wilderness*, 117.
52. Ibid., 118.
53. Ibid., 117.
54. Ibid., 105.
55. Ibid., 106.
56. Ibid., 303.
57. Douglas Bevington, *The Rebirth of Environmentalism: Grassroots Activism from the Spotted Owl to the Polar Bear* (Washington: Island Press, 2009), 29.
58. Turner, *The Promise of Wilderness*, 304.
59. Dave Foreman, *Ecodefense: A Field Guide to Monkeywrenching* (Tucson, AZ: Earth First! Books, 1985), 17.
60. Berlau, *Eco-Freaks*, 211.
61. Quoted in Neil Hrab, "Greenpeace, Earth First!, PETA: Radical Fringe Tactics Move Toward Center Stage," *Organizational Trends*, March 2004.

Retrieved from: http://capitalresearch.org/pubs/pdf/03_04_OT.pdf
62. Quoted in Cronon, *Uncommon Ground,* 1996, 83.
63. Ibid., 83–84.
64. Ibid., 84.
65. James M. Turner, "The Specter of Environmentalism: Wilderness, Environmental Politics, and the Evolution of the New Right," *Journal of American History* 96 (2009): 123–148, 130.
66. Turner, *The Promise of Wilderness,* 188.
67. Opposition to development in one's backyard is sometimes referred to as NIMBYism, standing for Not-in-my-backyard.
68. Turner, *The Promise of Wilderness,* 190.
69. Ibid., 192.
70. Ibid., 193.
71. Ibid., 192.
72. Turner, "The Specter of Environmentalism," 123.
73. Taylor and Klingle, "Environmentalism's Elitist Tinge."
74. Turner, "The Specter of Environmentalism," 125.
75. Ibid., 139.
76. Ibid.

Chapter 5

1. Ann Murray, *Smog Deaths in 1948 Led to Clean Air Laws,* NPR, 2009.
2. Geoffery Lean, "The Great Smog of London: The Air Was Thick with Apathy," *The Telegraph,* 2012.
3. Steve Connor, "Ice Pack Reveals Romans' Air Pollution," *The Independent,* 1994. Retrieved from: http://www.independent.co.uk/news/uk/ice-pack-reveals-romans-air-pollution-1450572.html
4. Indur Goklany, *Clearing the Air* (Washington, DC: The Cato Institute, 1999), 9.
5. Ibid., 11.
6. Ibid.
7. Ibid., 13.
8. Ibid.
9. Ibid., 25.
10. Adler, "The Fable of Federal Environmental Regulation."
11. Goklany, *Clearing the Air,* 26.
12. Bureau of Mines, *Ringelmann Smoke Chart,* U.S. Department of Interior, 1967.
13. Goklany, *Clearing the Air,* 17.
14. Ibid., 22.
15. Rachelle Oblack, *What Is Smog?*
16. Goklany, *Clearing the Air,* 27.
17. *Crankcase Emission Control.* n.d.
18. Goklany, *Clearing the Air,* 27.
19. Environmental Protection Agency, *Overview of EPA Authorities for Natural Resource Managers Developing Aquatic Invasive Species Rapid Response and Management Plans: CWA Section 404—Permits to Discharge Dredged or Fill Material,* 2013.
20. Ibid.
21. Goklany, *Clearing the Air.*
22. Ibid.
23. Goklany, *Clearing the Air,* 30.
24. Terry L. Anderson, ed. *Political Environmentalism* (Stanford, CA: Hoover Institution Press, 2000), 278.
25. Ibid., 283.
26. EPA, *Overview of EPA Authorities.*
27. Ibid., 285.
28. EPA, *Overview of EPA Authorities.*
29. Ibid., 288.
30. Goklany, *Clearing the Air,* 36.
31. Anderson, *Political Environmentalism,* 292–94.
32. Environmental Protection Agency, *The Clean Air Act in a Nutshell: How It Works.*

33. Andrew Morriss, "The Politics of the Clean Air Act," in *Political Environmentalism,* ed. Terry L. Anderson (Stanford: Hoover Institution Press, 2000), 298–299.
34. Bruce A. Ackerman and William T. Hassler, *Clean Coal, Dirty Air.* New Haven and London: Yale University Press, 1981).
35. Morriss, "The Politics of the Clean Air Act,"303.
36. This chapter makes no attempt to argue for or against the validity of global warming or climate change, but only includes the topic for its relevance to the case study.
37. U.S. Supreme Court Media, *Massachusetts v. Environmental Protection Agency,* Oyez: IIT Chicago-Kent College of Law, 2013.
38. Ibid.
39. *Massachusetts et al. v. EPA et al.,* 2007, 549 U.S. 497; 127 S. Ct. 1438; 167 L. Ed. 2d 248; 2007 U.S. LEXIS 3785; 75 U.S.L.W. 4149; 63 ERC (BNA) 2057; 37 ELR 20075; 20 Fla. L. Weekly Fed. S 128.
40. Nathan D. Riccardi, Necessarily Hypocritical: The Legal Viability of EPA's Regulation of Stationary Source Greenhouse Gas Emissions Under the Clean Air Act. *Boston College Environmental Affairs Law Review* 39(1)(2012): 213–241, 214.
41. U.S. Supreme Court Media, *Massachusetts v. Environmental Protection Agency.*
42. Ibid.
43. *Massachusetts et al. v. EPA et al.* 2007.
44. J. Austin, "*Massachusetts v. Environmental Protection Agency*: Global Warming, Standing and the US Supreme Court," *Review Of European Community & International Environmental Law, 16*(3)(2007): 368–371. doi:10.1111/j.1467-9388.2007.00567.
45. *Massachusetts et al. v. EPA et al.,* 2007.
46. Ibid.
47. Ibid.
48. Ibid.
49. Ibid.
50. Ibid.
51. Ibid.
52. Ibid.
53. Ibid.
54. Ibid.
55. Ibid.
56. Ibid.
57. Ibid.
58. Ibid.
59. Riccardi, *Necessarily Hypocritical.*
60. EPA, *Vehicle Standards and Regulations,* 2013.
61. EPA, *Endangerment and Cause or Contribute Findings for Greenhouse Gases under Section 202(a) of the Clean Air Act,* 2009.
62. EPA. *Prevention of Significant Deterioration and Title V Greenhouse Gas Tailoring Rule,* 2010.
63. Ibid.
64. Ibid.
65. R. Bravender, "EPA Issues Final 'Tailoring' Rule for Greenhouse Gas Emissions," *New York Times—Breaking News, World News & Multimedia,* May 13, 2010.
66. Riccardi, *Necessarily Hypocritical,* 229.
67. *Chevron U.S.A., Inc. v. Natural Resources Defense Council, Inc.,* 467 U.S. 837
68. Riccardi, *Necessarily Hypocritical,* 230.
69. Ibid., 231.
70. Ibid.
71. Friedrich Hayek, "The Use of Knowledge in Society," *Individualism and Economic Order* (1948), 77–91. Quote on 78.
72. Ibid., 79.

73. Angelo M. Codevilla, "Scientific Pretense vs. Democracy," *The American Spectator,* April 2009.
74. Ibid.
75. Daniel Kahneman, *Thinking, Fast and Slow* (New York: Farrar, Strauss, and Giroux, 2011).
76. Ibid., 201.
77. Ken Sexton, "Science and Policy in Regulatory Decision Making: Getting the Facts Right about Hazardous Air Pollutants," *Environmental Health Perspectives,* 103(1995): 213, 222, 213.
78. Ibid., 216.
79. Anne E. Smith, *An Evaluation of the $PM_{2.5}$ Health Benefits Estimates in Regulatory Impact Analyses for Recent Air Regulations,*" NERA Economic Consulting, December 2011.
80. Tony Cox, "Reassessing the Human Health Benefits from Cleaner Air," George Mason University, August 2011; Kathleen Hartnett White, May 2012. "EPA's Pretense of Science: Regulating Phantom Risks." Texas Public Policy Foundation, May 2012.
81. Geoffrey Kabat, "Over Reaching California EPA Regulators Promoted a False Breast Cancer Link," *Forbes Magazine,* October 2013.
82. Kahneman, *Thinking, Fast and Slow,* 413.
83. Ibid., 418.
84. William F. Pederson, Jr., "Why the Clean Air Act Works Badly," *University of Pennsylvania Law Review* 129 (May 1981): 1059, 1109, quote on 1059.
85. Andrew Morriss, "The Politics of the Clean Air Act," in *Political Environmentalism,* ed. Terry L. Anderson (Stanford, CA: Hoover Institution Press, 2000), 282.
86. Ibid., 265.
87. International Center for Technology Assessment, "Petition for Rulemaking and Collateral Relief Seeking the Regulation of Greenhouse Gas Emissions from New Motor Vehicles Under 202 of the Clean Air Act," October 1999.
88. Ibid.
89. Jonathan H. Adler, "Clean Politics, Dirty Profits," in *Political Environmentalism,* ed. Terry L. Anderson (Stanford, CA: Hoover Institution Press, 2000).
90. Ibid.
91. Ibid.
92. Ibid.
93. U.S. Congress. 1977. *Congressional Record.* 95th Congress, 1st Session, 1977, 18, 494.
94. Adler, *Clean Politics, Dirty Profits,* 8.
95. Quoted in Adler, *Clean Politics, Dirty Profits,* 3.
96. Michael Greenstone, "The Impacts of Environmental Regulations on Industrial Activity: Evidence from the 1970 and 1977 Clean Air Act Amendments and the Census of Manufactures," *Journal of Political Economy,* 110 (December 2002): 1175, 1219.
97. Cusick, Marie, "Corbett Calls Obama's Climate Change Proposal a War on Coal and Jobs," *State Impact,* June 25, 2013. Retrieved from: http://stateimpact.npr.org/pennsylvania/2013/06/25/corbett-calls-obamas-climate-change-proposal-a-war-on-coal-and-jobs/

Chapter 6

1. J. Matthew Haws, "Analysis Paralysis: Rethinking the Court's Role in Evaluating EIS Reasonable Alternatives," *University of Illinois Law Review* (2012): 537–576, 538.
2. Ibid.
3. 42 U.S.C Section 4321.
4. Haws, "Analysis Paralysis," 540; Daniel R. Mandelker, "The National

Environmental Policy Act: A Review of Its Experience and Problems," *Washington University Journal of Law and Policy* 32 (2010): 293–312, 293; Trevor Salter, "NEPA and Renewable Energy: Realizing the Most Environmental Benefit in the Quickest Time," *Environs* 34 (2011): 173–187, 176; Sam Kalen, "The Devolution of NEPA: How the APA Transformed the Nation's Environmental Policy." *William & Mary Environmental Law and Policy Review* 33 (2009): 483–548, 484, 545.

5. D. W. Schindler, "The Impact Statement Boondoggle," *Science,* May 7, 1976, 509; Linda Luther, *The National Environmental Policy Act: Background and Implementation,* CRS Report for Congress, November 16, 2005, 28.; Salter, "NEPA and Renewable Energy," 175; Jonathan DuHamel, "How NEPA Crushes Productivity," *Tucson Citizen,* May 6, 2013; J. Vines, S. Salek, and K. Desloover, "Reforming NEPA Review of Energy Projects," *King & Spalding Energy Newsletter,* December 2012.

6. Peter Alfano, "NEPA at 40: Procedure or Substance." Environmental Law Institute and The Council on Environmental Quality, 2010, 9–17; Richard Fristik, *Mitigation Under NEPA: Failed Promises?* Prepared for U.S. Department of Agriculture, November 2012, 2; Salter, "NEPA and Renewable Energy," 177.

7. Luther, *NEPA: Background.*
8. (42 U.S.C Sections 4321–4347)
9. (*C.F.R.*, 1978, Title 40, Part 1501.3)
10. (*C.F.R.*, 1978, Title 40, Section 1502.14)
11. Alfano, "NEPA at 40," 9–10; Haws, "Analysis Paralysis," 54; Salter, "NEPA and Renewable Energy," 177; Kalen, "The Devolution of NEPA," 484.
12. Haws, "Analysis Paralysis," 544.
13. House Committee on Natural Resources, *NEPA: Lessons Learned and Next Steps,* House of Representatives Hearing, November 17, 2005, U.S. Government Printing Office.
14. U.S. Department of Energy, *National Environmental Policy Act: Lessons Learned,* September 6, 2013.
15. Ibid.
16. Salter, "NEPA and Renewable Energy."
17. DOE, *NEPA: Lessons Learned.*
18. House Committee on Natural Resources, *NEPA: Lessons Learned and Next Steps.*
19. Steve Pociask and Joseph Fuhr, *Progress Denied: A Study on the Potential Economic Impact of Permitting Challenges Facing Proposed Energy Projects,* March 10, 2011, 2. Prepared for U.S. Chamber of Commerce.
20. Diane Katz and Craig Manson, "The National Environmental Policy Act," *The Heritage Foundation: Leadership for America,* 2012.
21. Mandelker, "NEPA: A Review," 294.
22. Ibid.
23. Katz and Manson, "NEPA."
24. Katz and Manson, "NEPA"; Mandelker, "NEPA: A Review"; Salter, "NEPA and Renewable Energy"; Vines et al., "Reforming NEPA."
25. Marris, *Rambunctious Garden.*
26. Mandelker, "NEPA: A Review."
27. Vines et al., "Reforming NEPA."
28. Enviro Defenders. *Legal Handbook for Environmental Activists: NEPA & CEQA.*
29. Katz and Manson, "NEPA."
30. Committee and Natural Resources, *NEPA: Lessons Learned.*
31. Fran Hoffinger, "Environmental Impact Statements: Instruments for Environmental Protection or Endless Litigation?" *Fordham Urban Law Journal* 11 (1982): 527–566.

32. Michael Coulter, *House Task Force Hears Testimony on Improving Decades-Old Environmental Law*. The Heartland Institute, August 1, 2005.
33. Haws, "Analysis Paralysis."
34. Hoffinger, "Environmental Impact Statements," 556; William Murray Tabb, "The Role of Controversy in NEPA: Reconciling Public Veto with Public Participation in Environmental Decisionmaking," *William & Mary Environmental Law and Policy Review* 21 (1997): 175–231, 220; Haws, "Analysis Paralysis," 547.
35. Katz and Manson, "NEPA."
36. Dinah Bear, "Some Modest Suggestions for Improving Implementation of the National Environmental Policy Act," *Natural Resources Journal*, 2013, 931–960, 932.
37. Charles H. Eccleston, *The EIS Book: Managing and Preparing Environmental Impact Statements*. (Boca Raton: Florida, CRC Press, 2014) 268.
38. Susan M. Smillie and Lucinda Low Swartz, *Achieving the 150-Page Environmental Impact Statement*, 2002.
39. Ibid.
40. Ibid.
41. Quoted in House Committee on Natural Resources, *NEPA: Lessons Learned*.
42. Council on Environmental Quality, *Modernizing NEPA Implementation*, U.S. Executive Office of the President. 2003, 121.
43. Haws, "Analysis Paralysis."
44. (Executive Order 13604)
45. Haws, "Analysis Paralysis" 575.
46. Tony Allen, Personal conversation between Tony Allen, formerly of Symbiotics, Inc. and Kenneth Sim regarding the City of Afton, Wyominig Micro-Hydro Project on Swift Creek, 2013.
47. Fristik, *Mitigation Under NEPA*.
48. Ibid.
49. Ibid.
50. (40 C.F.R. 1508.4)
51. Tabb, "The Role of Controversy."
52. Robert Smythe and Caroline Isber, "NEPA in the Agencies: A Critique of Current Practices," *Environmental Practice*, 5(4) (2003): 290–297.
53. Bear, "Modest Suggestions."
54. Haws, "Analysis Paralysis."
55. Salter, "NEPA and Renewable Energy."
56. Vines et al., "Reforming NEPA."
57. Ibid.
58. Haws, "Analysis Paralysis."
59. Coulter, *House Task Force Hears Testimony*.
60. Vines et al., "Reforming NEPA."
61. Haws, "Analysis Paralysis."
62. Coulter, *House Task Force Hears Testimony*.
63. Hoffinger, "Environmental Impact Statements."
64. Luther, *NEPA: Background*.
65. Ron Bass, *Congressional Task Force Recommends Changes to the National Environmental Policy Act (NEPA)*, ICF International.
66. Ibid.
67. Salter, "NEPA and Renewable Energy"; W. Rein, E. Andreas, and K. Kennedy, "NEPA and Renewable Energy Practices: Streamlining Sustainability," *Association of Corporate Counsel*, June 28, 2012.
68. Salter, "NEPA and Renewable Energy."
69. Western Energy Alliance, *National Environmental Policy Act: Government Delays Preventing Jobs and Economic Growth*. Retrieved June 23, 2015 from http://www.westernenergyalliance.org/knowledge-center/land/onshore-development/national-environmental-policy-act-nepa
70. Ibid.

71. House Committee on Natural Resources, *NEPA: Lessons Learned.*
72. Vines et al., "Reforming NEPA."

Chapter 7

1. Case Western Reserve University, "Cuyahoga River Fire," *The Encyclopedia of Cleveland History,* June 25, 1997.
2. "America's Sewage System and the Price of Optimism," *Time,* August 1, 1969.
3. Ann Powers, *Federal Water Pollution Control Act (1948) (Major Acts of Congress),* 2004.
4. Jonathan Adler, "Fables of the Cuyahoga: Reconstructing a History of Environmental Protection," *Fordham Environmental Law Journal* 14 (2002): 89–146, 91.
5. Ibid., 105.
6. Russell McLendon, "Clean Water Act Is 40 Years Old: Landmark Water Law Hits a Milestone During Critical Time," *The Huffington Post,* September 11, 2012.
7. UC Santa Barbara Department of Geography, *Keith Clarke Quoted by the BBC Re Offshore Drilling,* May 27, 2010.
8. McLendon, "Clean Water Act is 40."
9. Committee on Environment and Public Works, US Senate, *Senate Report 111–361: Clean Water Restoration Act,* December 10, 2010.
10. Drew Caputo, "A Job Half Finished: The Clean Water Act After Twenty-Five Years," *The Environmental Law Reporter* 27 (November 1997): 10574.
11. Region 6 Office, Environmental Protection Agency. 2011, October 4. *Clean Water Act,* October 4, 2011; Robert B. Semple, "Happy Birthday, Clean Water Act," *New York Times Opinion Pages,* October 16, 2012.
12. Victor B. Flatt, "Dirty River Runs Through It (The Failure of Enforcement in the Clean Water Act)," *Boston College Environmental Affairs Law Review* 25 (September 1, 1997): 1–45, 2–4.
13. Adler, "Fables of the Cuyahoga," 111–112; General Accounting Office, *Water Pollution Abatement Program: Assessment of Federal And State Enforcement Efforts,* March 23, 1972.
14. Region 6, EPA, *Clean Water Act.*
15. Caputo, "A Job Half Finished"; Flatt, "Dirty River," 11; Environmental Protection Agency, *National Pollutant Discharge Elimination System (NPDES),* March 12, 2009.
16. Flatt, "Dirty River," 12.
17. Ibid.
18. Robert McClure and Bonnie Stewart, *Clean Water Act's Anti-Pollution Goals Prove Elusive,* Oregon Public Broadcasting, July 18, 2012. Retrieved from: http://earthfix.opb.org/water/article/anti-pollution-goals-elude-clean-water-act-enforce/
19. Flatt, "Dirty River," 8.
20. Ibid.
21. Environmental Protection Agency, *What is Nonpoint Source Pollution?* August 27, 2012. Retrieved from: http://water.epa.gov/polwaste/nps/whatis.cfm
22. Flatt, "Dirty River," 12.
23. Ibid.
24. Ibid.
25. Ibid., 2–4.
26. Peter Hill and Roger E. Meiners, *Who Owns The Environment?* (Lanham, MD: Rowman & Littlefield, 1998), 100.
27. Ibid., 87.
28. Flatt, "Dirty River," 2–4.
29. McLendon, "Clean Water Act Is 40."
30. Quoted in Ibid.
31. Ibid.

32. Office of Wetlands, Oceans, & Watersheds, Environmental Protection Agency, *A National Evaluation of the Clean Water Act Section 319 Program,* November 2011, 4.
33. McLendon, "Clean Water Act Is 40."
34. Fish and Wildlife Service, Department of the Interior, *Digest of Federal Resource Laws,* January 10, 2013.
35. Ibid.
36. Claudia Copeland, *Clean Water Act: A Summary of the Law,* prepared for Congressional Research Service, April 23, 2010, 1.
37. Roger E. Meiners and Bruce Yandle, "How the Common Law Protects the Environment: Curbing Pollution—Case-By-Case," *PERC Reports* 16 (June 1998): 7–9.
38. Hill and Meiners, *Who Owns,* 87.
39. Ibid., 87–114.
40. Meiners and Yandle, "How the Common Law Protects."
41. Hill and Meiners, "Who Owns?" 110.
42. Meiners and Yandle, "How the Common Law Protects," 7.
43. Ibid.
44. Ibid., 8.
45. Ibid., 7.
46. Committee on Environment and Public Works, *Senate Report 111-361.*
47. Copeland, *Clean Water Act.*
48. McLendon, *Clean Water Act is 40.*
49. Office of Wetlands, *A National Evaluation,* 4.
50. Ibid., 1.
51. Environmental Protection Agency, *Chesapeake Bay TMDL Executive Summary,* December 29, 2010.
52. Ibid.
53. Ibid.
54. Karl Blankenship, "Local Officials Worry That TMDL Actions Are Much Too Costly," *Bay Journal,* July 1, 2011.
55. Natural Resources Conservation Service, U.S. Department of Agriculture. *Assessment of the Effects of Conservation Practices on Cultivated Cropland in the Chesapeake Bay Region,* March 2011.
56. Ibid.
57. Ibid.
58. S. Robinson, *Comments on the EPA Draft Chesapeake Bay TMDL,* National Association of Conservation Districts, 2010.
59. Ibid.
60. Blankenship, "Local Officials Worry."
61. Ibid.
62. Quoted in Ibid.
63. Ibid.
64. E. B. Ferguson, "Farm Lobby Likely to Appeal Chesapeake Bay Pollution Diet Ruling," *Capital Gazette,* September 24, 2013.
65. Ibid.
66. Environmental Protection Agency, 2013(g), March 12. *Clean Water Act (CWA).*
67. Coulter, *House Task Force Hears Testimony.*
68. Ibid., 17.
69. William N. Hines, "Nor Any Drop to Drink: Public Regulation of Water Quality Part I: State Pollution Control Programs," *Iowa Law Review* 52 (1966): 186–235, 215.
70. Government Accountability Office (GAO). *Clean Water Act: Longstanding Issues Impact EPA's and States' Enforcement Efforts,* October 15, 2009, 4. Retrieved from: http://www.gao.gov/new.items/d10165t.pdf.
71. Adler, "Fables of the Cuyahoga," 102–103.
72. Wallace E. Oates, *A Reconsideration of Environmental Federalism,* Resources for the Future, November 2001, 16.
73. Ibid.
74. Government Accountability Office (GAO). *Clean Water Act: Longstanding Issues Impact EPA's and States' Enforcement Efforts,* October 15, 2009, 4.

75. Ibid., 5.
76. Ibid., 4.
77. Ibid.
78. D. Engelberg, K. Butler, C. Brunton, A. Chirigotis, J. Hamann, and M. Reed, *EPA Must Improve Oversight of State Enforcement,* prepared for U.S. Environmental Protection Agency, Office of Inspector General, December 9, 2011, 3.
79. Ibid., 15.
80. Ibid., 16.
81. Sean Doogan, *Both State, Feds Probe August EPA Task Force Raids near Chicken. Alaska Dispatch,* October 10, 2013.
82. Ibid.

Chapter 8

1. Oregon Department of Forestry. (n.d.(a)). History and Legal Mandates of Oregon State Forests. Oregon.gov. Retrieved from http://www.oregon.gov/odf/pages/state_forests/history.aspx.
2. Oregon Department of State Lands. (2011, November). Elliott State Forest Management Plan: Chapter 2. Oregon.gov. Retrieved from http://www.oregon.gov/odf/state_forests/docs/esf/elliott_fmp_2011/elliott sf_2011_fmp_final.pdf.
3. Caro, *Conservation by Proxy;* L. Scott Mills, Michael E. Soulé, and Daniel F. Doak, "The Keystone-Species Concept in Ecology and Conservation," *Bioscience* 43 (1993): 219–224.
4. Caro, *Conservation by Proxy.*
5. Justin Wright, Clive G. Jones, and Alexander G. Flecker, "An Ecosystem Engineer, the Beaver, Increases Species Richness at the Landscape Scale," *Oecologia* 132 (1)(2002): 96–101.
6. Ronald L. Ives, "The Beaver-Meadow Complex," *Journal of Geomorphology* 5 (1942): 191–203.
7. Land and Water Conservation Fund Act of 1965, Public Law 88-578 Title 16, United States Code.
8. Public Law 89-669, United States Code.
9. Taking has a particular meaning in the ESA; it is defined as "to harass, harm, pursue, hunt, shoot, wound, kill, trap, capture, or collect, or to attempt to engage in any such conduct" (U.S. Government Printing Office, n.d.(b), 1,784).
10. J. A. Michael, "The Endangered Species Act and Private Landowner Incentives," *United States Department of Agriculture: Animal and Plant Health Inspection Service,* August 2000, 29.
11. Ibid.
12. Norm Dicks, "Statement of Congressman Norm Dicks re Hatchery Salmon," *Project Vote Smart,* May 20, 2004.
13. "The History of the Endangered Species Act," *The Thoreau Institute.*
14. National Oceanic and Atmospheric Administration, *Full Text of the Endangered Species Act (ESA).*
15. U.S. Government Printing Office, 16 U.S.C. § 1532, 1784.
16. Ibid.
17. U.S. Fish and Wildlife Service, *ESA Basics: 40 Years of Conserving Endangered Species,* January 2013, 1.
18. U.S. Government Printing Office, 16 U.S.C. § 1532, 1784.
19. U.S. Fish and Wildlife Service, *Our Endangered Species Program and How It Works with Landowners,* June 2009, 1. Retrieved from: http://www.fws.gov/endangered/esa-library/pdf/landowners.pdf
20. U.S. Fish and Wildlife Service. 2013, July(c). *1982 ESA Amendment,* 1. Retrieved from: http://www.fws.gov/endangered/laws-policies/esa-1982.html

21. *Tennessee Valley Authority (TVA) v. Hill et al.*, 437 U.S. 153 (1978).
22. Tennessee Valley Authority (TVA), *Frequently Asked Questions About TVA*.
23. Z. J. B. Plater, "In the Wake of the Snail Darter: An Environmental Law Paradigm and Its Consequences," *Journal of Law Reform* 19 (July 1, 1986), 813.
24. R. Wilson, "Tellico Dam Still Generating Debate," *Knoxville News Sentinel*, April 13, 2008.
25. Plater, "In the Wake of the Snail Darter," 810.
26. See for a detailed description of the National Environmental Policy Act.
27. Plater, "In the Wake of the Snail Darter."
28. Ibid.
29. Zygmunt J. B. Plater, *The Snail Darter and the Dam: How Pork-Barrel Politics Endangered a Little Fish and Killed a River* (New Haven, CT: Yale University Press, 2013).
30. Plater, "In the Wake of the Snail Darter."
31. *TVA v. Hill*, 1978.
32. Plater, "In the Wake of the Snail Darter."
33. Ibid.
34. According to Plater, a "half-hearted veto threat" was made by President Carter, and the Cherokee Indians, to whom the area flooded was sacred tribal lands from which they'd been driven "in a forced emigration, which culminated in the Trail of Tears to Oklahoma." Quoted in Plater, "In the Wake of the Snail Darter."
35. *Babbitt, Secretary of Interior v. Sweet Home Chapter of Communities for a Great Oregon*, 515 U.S. 687 (1995).
36. Recall that the ESA set an extremely broad definition of "take" as "to harass, harm, pursue, hunt, shoot, wound, kill, trap, capture, or collect, or to attempt to engage in any such conduct" (U.S. Government Printing Office, n.d.(b), 1,784).
37. Cornell University Law School, *Babbitt v. Sweet Home Chapt. Comms. for Ore.*, 515 U.S. 687 (1995).
38. Ibid.
39. *Chevron U.S.A., Inc. v. Natural Resources Defense Council, Inc.* (Chevron v. NRDC), *Legal Information Institute*. Retrieved from: http://www.law.cornell.edu/supct/html/historics/USSC_CR_0467_0837_ZO.html
40. These terms, while disparate in meaning, are frequently used interchangeably to refer to the planetary climate shifts attributed to human released emissions of greenhouse gases; the term "climate change" will be used here to refer to the phenomenon.
41. S. J. Morath, "Endangered Species Act: A New Avenue for Climate Change Litigation." *Public Land & Resources Law Review* 29 (2008): 23–40. Retrieved from http://scholarship.law.umt.edu/cgi/viewcontent.cgi?article=1253&context=plrlr.
42. National Wildlife Federation, *Keeping the Endangered Species Act Strong*
43. M. C. O'Connor, "Can the Endangered Species Act Protect Against Climate Change?" *Outside,* February 28, 2013.
44. The suggestion is not that the benchmarks are any more readily achievable, simply that scientists can determine with a fair degree of confidence historical atmospheric levels of gases from ice samples and, therefore, at least have some definitive benchmark state by which success can be measured.
45. Arcata Fish and Wildlife Office, U.S. Fish and Wildlife Service, *Northern*

Spotted Owl Species Profile, July 5, 2011. Retrieved from: http://www.fws.gov/arcata/es/birds/nso/ns_owl.html; Defenders of Wildlife, "Northern Spotted Owl." Retrieved from http://www.defenders.org/northern-spotted-owl/basic-facts
46. Oregon Fish and Wildlife Office, U.S. Fish and Wildlife Service, *Northern Spotted Owl,* November 5, 2012. Retrieved from: www.fws.gov/oregonfwo/species/data/northernspottedowl/
47. Oregon Fish and Wildlife Office, *Northern Spotted Owl;* Defenders of Wildlife, "Northern Spotted Owl."
48. Cornell Lab of Ornithology, *Barred Owl*. Retrieved from http://www.allaboutbirds.org/guide/barred_owl/id; M.A. Johnson, "Feds Move Ahead with Plans to Kill Barred Owls to Save Spotted Owls," *U.S. News,* July 23, 2013.
49. Cornell Lab of Ornithology, *Barred Owl*.
50. Oregon Fish and Wildlife Office, *Northern Spotted Owl*.
51. Oregon Fish and Wildlife Office, *Northern Spotted Owl;* K. B. Livezey, "Killing Barred Owls to Help Spotted Owls I: A Global Perspective," *Northwestern Naturalist* 91(2)(2010), 107–133.
52. Livezey, "Killing Barred Owls I."
53. Johnson, "Feds Move Ahead"; Livezey, "Killing Barred Owls I." Fish and Wildlife Service has killed sea lions and cormorants to protect salmon runs, but killing more than 3000 members of a species is unprecedented.
54. Johnson, "Feds Move Ahead"; Livezey, "Killing Barred Owls I"; Associated Press, "Feds May Wage War on Owls to Save Others," *NBC News,* June 15, 2005.
55. Johnson, "Feds Move Ahead."
56. Ibid.
57. Ibid.
58. S. Learn, "Northern Spotted Owl Marks 20 Years on Endangered Species List," *OregonLive.com,* February 5, 2001.
59. Quoted in Jonathan Adler, "Anti-Conservation Efforts," *Regulation* (Winter 2009), 54.
60. Adler, "Anti-Conservation Efforts." Ibid.
61. D. Lueck and J. A. Michael, 2003. Preemptive Habitat Destruction under the Endangered Species Act. *Journal of Law and Economics* 46(1) (2003), 27–60.
62. Nature Conservancy. "Mississippi Red-Cockaded Woodpecker," February 14, 2011.
63. Adler, "Anti-Conservation Efforts," 23.
64. Lueck and Michael, "Pre-emptive Habitat Destruction," 53; M. T. O'Keefe, "Red-Cockaded Woodpeckers," *Florida Wildlife Viewing*.
65. R. O'Toole, "Census Bureau: 94.6% of U.S. Is Rural Open Space," *Heartlander Magazine,* July 1, 2003; SummitPost, "Public and Private Land Percentages by US States."
66. Endangered Species Act: Information on Species Protection on Nonfederal Lands. (1994, December 20). Retrieved June 5, 2015, from http://www.gao.gov/products/RCED-95-16
67. Lynn Dwyer, Dennis Murphy, and Paul Ehrlich, "Property Rights Case Law and the Challenge to the Endangered Species Act," *Conservation Biology* 9(4) (1995): 725–741, 736.
68. Larry McKinney, "Reauthorizing the Endangered Species Act—Incentives for Rural Landowners", in *Building Economic Incentives into the Endangered Species Act: A Special Report from*

Defenders of Wildlife, ed. Wendy E. Hudson (Washington, DC: Defenders of Wildlife, 1993), 63–65.
69. David S. Wilcove, Michael J. Bean, Robert Bonnie, and Margaret McMillan, "Rebuilding the Ark, Toward a More Effective Endangered Species Act for Private Land," December 5, 1996, *On-Line Learning at the Cumberland School of Law*.
70. Ibid., 6.
71. This should not be confused with the definition of "taking" under the Endangered Species Act.
72. L. A. Welch, 1994. "Property Rights Conflicts Under the Endangered Species Act: Protection of the Red Cockaded Woodpecker," in *Land Rights: The 1990's Property Rights Rebellion*, ed. B. Yandle (Lanham, MD: Rowman and Littlefield, 1994), 186.
73. Dwyer et al., "Property Rights Case Law."
74. Ibid.
75. *Lucas v. South Carolina Coastal Council*, 505 U.S. 1003 (1992).
76. *Dolan v. City of Tigard*, 512 U.S. 374 (1994).
77. Quoted in Lueck and Michael, "Preemptive Habitat Destruction," 27.
78. Reason Foundation. Bald Eagle Off Endangered List In Spite of Feds, Not Because of Them, June 27, 2007.
79. Big Cat Sticker Shack, "Shoot, Shovel & Shut up" (Bumper Sticker), *Amazon*; CafePress, Anti Wolf Gifts & Merchandise; No Wolves, Anti Wolf Tshirt.
80. Legal Ruralism. "Shoot, Shovel, and Shut up," *Legal Ruralism,* November 10, 2011.
81. L. Pritchett, "Sight the Gun High," *Natural Resources Journal* 46(1) (2006): 1–8.
82. R. Bailey, Shoot, Shovel, and Shut Up, December 31, 2003; T. Dewey and J. Smith, "Canis Lupus," *Animal Diversity Web,* 2002.
83. A. Rosenberg, "Wolves, the Endangered Species Act, and Why Scientific Integrity Matters," *The Equation,* August 19, 2013.
84. J. Morrison, "Shoot, Shovel & Shut-Up," *NewsWithViews,* August 14, 2004; D. Strain, "House Strikes Proposed Ban on Endangered Species Listings," *ScienceInsider*. July 28, 2011.
85. Strain, "House Strikes Proposed Ban."
86. Ibid.
87. Ibid.
88. R. McLendon, "Endangered Species Act Stirs New Debates," *Mother Nature Network,* March 13, 2012.
89. Associated Press, "Feds May Wage War."
90. J. Fong, "Fox Sees Conspiracy in Effort to Protect Lizard Species," *Media Matters,* April 29, 2011; D. Hastings, "Excessive Endangered Species Act Litigation Threatens Species Recovery, Job Creation and Economic Growth," *House Committee on Natural Resources,* December 6, 2011.
91. "Bladderpod Controversy Reinforces Need to Reform Endangered Species Act, *Tri-City Herald,* August 11, 2013.
92. M. Gerhart, "Climate Change and the Endangered Species Act," *Ecology Law Quarterly* 36 (2009): 167; A. C. Revkin, "U.S. Curbs Use of Species Act in Protecting Polar Bear," *New York Times,* May 8, 2009.
93. Rural Liberty Alliance, "The ESA: How Litigation Is Costing Jobs and Impeding Recovery," December 6, 2011.
94. Hastings, "Excessive Endangered Species Act Litigation."
95. Center for Biological Diversity, *Defending Endangered Species.*; Defenders of Wildlife, *Endangered Species Act 101.*
96. Natural Resources Defense Council, *Court Upholds Endangered Species Act*

Protection for Polar Bears. M. Clayton, "New Tool to Fight Global Warming: Endangered Species Act?" *The Christian Science Monitor,* September 7, 2007.
97. Ridenour, "TESRA."
98. Quoted in E. Gies, "Report: Endangered Species Act Works," *Forbes,* May 30, 2012.
99. A. Wetzler, "Should We Pay Landowners Not to Kill Endangered Species?" *Natural Resources Defense Council,* August 24, 2007.
100. Ridenour, "TESRA"; Adler, "Anti Conservation Measures."
101. B. Seasholes, "Bald Eagle Off Endangered List In Spite of Feds, Not Because of Them," *Reason Foundation,* June 27, 2007; "The Bald Eagle, DDT, and the Endangered Species Act: Examining the Bald Eagle's Recovery in the Contiguous 48 States," *Reason Foundation,* June 1, 2007.
102. Ridenour, "TESRA."
103. Ibid.
104. Seasholes, "The Bald Eagle, DDT, and the ESA"; Ridenour, "TESRA"; K. Suckling, N. Greenwald, and T. Curry, "On Time, On Target How the Endangered Species Act is Saving America's Wildlife," *Center for Biological Diversity.*
105. Center for Biological Diversity, *110 Success Stories for Endangered Species Day,* 2012.
106. K. Suckling, Testimony on "The Endangered Species Act: How Litigation Is Costing Jobs and Impeding True Recovery Efforts," *Center for Biological Diversity,* December 2, 2011, 1.
107. Suckling et al., "On Time, On Target."
108. R. Hopper, "Inflated Endangered Species Act 'Success Stories' Revealed," *PLF Liberty Blog,* June 5, 2012.
109. Eco-Now, "Endangered Species Act Success Stories," July 2, 2012; Gies, "Report: ESA Works"; "Is the Endangered Species Act a Success or Failure?" *Scientific American,* August 9, 2012; Hopper, "Inflated ESA 'Success Stories.'"
110. House Committee on Natural Resources, *Endangered Species Act.*
111. E. H. Buck, M. L. Corn, K. Alexander, A. Sheikh, and R. Meltz, "The Endangered Species Act (ESA) in the 112th Congress: Conflicting Values and Difficult Choices," *Federation of American Scientists,* June 13, 2012.
112. D. Hastings, "Hastings: Time to Improve the Endangered Species Act," *Washington Times,* May 18, 2012.
113. Ibid.
114. Ibid.

Chapter 9

1. Peter Kareiva, Michelle Marvier, and Robert Lalasz, *Conservation in the Anthropocene: Beyond Solitude and Fragility* (2012).
2. Stelle Slootmaker, "John Muir: Nature's Evangelist?" Grand Rapids Institute of Information Democracy, 2011. Retrieved from http://griid.org/2011/01/21/john-muir-natures-evangelist/
3. Kareiva et al., *Conservation in the Anthropocene*
4. Mark Wilton Harvey, *Wilderness Forever: Howard Zahniser and the Path to the Wilderness Act* (Seattle: University of Washington Press, 2005), ix.
5. Ibid., xi.
6. Ibid., xii.
7. Ibid., xiii.
8. The Wilderness Act, 1964. U.S. Code. Title 16, Sections 1131–1136.
9. Wilderness Act, 1964, Sections 1131–1136.

10. Famous Quotes. (n.d.). Retrieved June 5, 2015, from http://wilderness.org/article/famous-quotes
11. Ibid.
12. Anderson, *Political Environmentalism*, 22.
13. Anderson, *Political Environmentalism*, Ibid., 219.
14. Ibid., 221.
15. Ibid., 222.
16. Ibid., 224.
17. Ibid., 229.
18. Ibid., 239.
19. Craig Allin, "Understanding the Wilderness Act of 1964," *Cornell College*, 2002, 2.
20. Allin, "Understanding," Ibid., 2.
21. Allin, "Understanding," Ibid., 4.
22. Allin, "Understanding," Ibid., 4.
23. The Wilderness Act, 1964.
24. Southern Utah Wilderness Alliance, "About SUWA," n.d.
25. Southern Utah Wilderness Alliance, "Protecting Greater Canyonlands," n.d.
26. America's Red Rock Wilderness Act of 2009, S. 799. 111th Congress (2009). Retreived from https://www.govtrack.us/congress/bills/111/s799.
27. House Committee on Natural Resources, "Utah Congressional Delegation Expresses Bipartisan Opposition to America's Red Rock Wilderness Act," October 1, 2009.
28. House Committee on Natural Resources, "Utah Congressional Delegation." Ibid.
29. Dick Durbin, "Durbin and Hitchey Introduce Bill to Protect America's Red Rock Wilderness," n.d.
30. Bonner Cohen, "Red Rock Wilderness Bill," *Committee for a Constructive Society*, July 13, 2009.
31. David N. Cole, "Management Dilemmas That Will Shape Wilderness in the 21st Century," *Journal of Forestry* 99 (1) (January 2001): 4–8, 5.
32. U.S. Forest Service, United States Department of Agriculture, Wilderness Management 101, n.d.
33. Ted Stroll, "Aw, Wilderness!" *New York Times*, August 26, 2010.
34. Stroll, "Aw, Wilderness!" Ibid.
35. Stroll, "Aw, Wilderness!"
36. *Southern Utah Wilderness Alliance v. Norton*, 301 F.3d 1217 (2002).
37. The Wilderness Act, 1964.
38. *Southern Utah Wilderness Alliance v. Norton*, 301 F.3d 1217 (2002).
39. *Norton v. Southern Utah Wilderness Alliance*, 542 U.S. 55 (2004).
40. Randy T Simmons and Ryan M. Yonk, "Economics Impacts of Southern Utah Wilderness Alliance Litigation on Local Communities," Utah State University, 2013.
41. Simmons and Yonk, "Economic Impacts."
42. Southern Utah Wilderness Association, November 4, 2013. "Court Strikes Down Controversial BLM Land Use Plan." Retrieved from http://suwa.org/court-strikes-down-controversial-blm-land-use-plan/.
43. Laitos, Jan G., and Rachael B. Gamble. 2008. The Problem with Wilderness. Harvard Environmental Law Review, 32: 503–569, 510.
44. Harvey, *Wilderness Forever*.
45. Michael Frome, Michael. 2007. *Rebel on the Road: And Why I Was Never Neutral*. (Truman State University Press: Kirksville, MO, 2007).
46. Quoted in Marris, *Rambunctious Garden*, 19.
47. Harold Wood, "Quotations from John Muir," Sierra Club website, 2013.
48. William Badè, "Introduction to 'A Thousand-Mile Walk to the Gulf'," by John Muir, Sierra Club, 1916.

49. Ryan M. Yonk, Randy T Simmons, and Brian C. Steed. 2013. *Green vs. Green: The Political, Legal, and Administrative Pitfalls Facing Green Energy Production*, New York: Routledge.
50. Laitos and Gamble, "The Problem with Wilderness," 511.
51. Laitos and Gamble, "The Problem with Wilderness," 511.
52. Dennis Coates and Brad R. Humphreys, "The Economic Consequences of Professional Sports Lockouts and Strikes," *Southern Economic Journal* 67 (3) (January 2001), 737–747.
53. WildEarthGuardians website (Wildearthguardians.org,) n.d.
54. Ryan M. Yonk, Randy T Simmons, and Kayla Dawn Harris. "Economic Impacts of WildEarth Guardians on Local Communities." 2012 http://www.strata.org/wp-content/uploads/ipePublications/Economic-Impacts-of-WildEarth-Guardians-Litigation-on-Local-Communities.pdf
55. Clint Bolick, "Public Lands and Private Incentives," *The Freeman*, July 14, 1985.
56. Alpert, P. (n.d.). "Managing the wild: Should stewards be pilots?" Retrieved June 5, 2015, from leopold.wilderness.net/pubs/531_1.pdf
57. Cole, "Management Dilemmas."
58. Quoted in Alpert et al., "Managing the Wild."
59. Wilderness Watch, "Recent Issues." Retrieved June 22, 2015 from http://wildernesswatch.org/issues/
60. Wilderness Watch, "Recent Issues."
61. Wilderness Act, 1964.
62. Ryan M. Yonk, Randy T Simmons, and Kayla Dawn Harris. "Economic Impacts of Southern Utah Wilderness Alliance." http://www.strata.org/wp-content/uploads/ipePublications/Economic-Impacts-of-Southern-Utah-Wilderness-Alliance-on-Lcoal-Comminities.pdf
63. Ibid.
64. Ibid.
65. Charles Michael Ray,"Forest Near Mount Rushmore Suffers Beetle Attack," NPR, September 9, 2010.
66. Phil Davies, "The beetle and the damage done," Fedgazette, Federal Reserve Bank of Minneapolis, November, 2012.
67. Ibid.
68. Ibid.
69. Ray, "Forest Near Mount Rushmore."
70. *Transforming Forest Waste to Bio Fuels and the Renewable Fuels Standard: Field hearing before the Committee on Agriculture, Nutrition, and Forestry, United States Senate*, 110th Cong. 8(2008) (statement of Craig Bolzien, Forest Supervisor, Black Hills National Forest, Custer, South Dakota).
71. Ray, "Forest Near Mount Rushmore."
72. Black Hills National Forest. n.d. "Mountain Pine Beetle: Frequently Asked Questions." U.S. Department of Agriculture. Retrieved June 22, 2015 from: http://www.fs.usda.gov/Internet/FSE_DOCUMENTS/stelprdb5304956.pdf.
73. Wilderness Act, 1964.
74. *The Black Hills and the Mountain Pine Beetle*, video, 2011, South Dakota Public Broadcasting. Retrieved June 22, 2015from http://watch.sdpb.org/video/1977593380/.
75. Douglas W. Scott, "A Wilderness-Forever Future: A Short History of the National Wilderness Preservation System," National Park Service, June 2001, rev. 2003.
76. The Wilderness Act, 1964.
77. Allin, "Understanding the Wilderness Act," 17.

78. Quoted in Alpert et al., "Managing the Wild."
79. Cole, "Management Dilemmas," 5.

Chapter 10

1. Dan Bobkoff, "10 Years After the Blackout, How Had the Power Grid Changed?" *National Public Radio*, August 14, 2013.
2. James Barron, "Power Surge Blacks Out Northeast," *New York Times*, August 15, 2003.
3. Barron, "Power Surge"; Bobkoff, "10 Years Later."
4. U.S.-Canada Power System Outage Task Force, Final Report on the Implementation of the Task Force Recommendations, U.S. Department of Energy, September 2006,
5. Ibid.
6. Congressional Research Service, Energy Policy Act of 2005: Summary and Analysis of Enacted Provisions, March 8, 2006.
7. Ibid.
8. Ibid.
9. Database of State Incentives for Renewables & Efficiency, Federal Incentives/Policies for Renewables & Efficiency, 2011.
10. M. C. Lott, "Solyndra—Illuminating Energy Funding Flaws?" *Scientific American*, September 27, 2011.
11. Matthew L. Wald and Michael Kanellos, "F.B.I. Raids Solar Firm That Got U.S. Loans," *New York Times*, September 8, 2011.
12. Lott, "Solyndra."
13. Ibid.
14. David Biello, "Cylindrical Solar Cells Give a Whole New Meaning to Sunroof," *Scientific American*, October 7, 2008.
15. Ibid.
16. Ucilia Wang, "Solyndra Rolls Out Tube-Shaped Thin Film," *Greentechsolar*. October 7, 2008.
17. Ibid.
18. Biello, "Cylindrical Solar Cells."
19. Ibid.
20. Ibid.
21. U.S. House of Representatives, *The Solyndra Failure: Majority Staff Report*, August 2, 2012, 10–15.
22. D'Angelo Gore and Eugene Kiely, "Obama's Solyndra problem," *FactCheck.org*, October 7, 2011.
23. Ibid.
24. Brad Plumer, "Energy Secretary Chu Felt Pressure to Speed Up Loans to Firms Like Solyndra. *Washington Post*, September 30, 2011.
25. U.S. House of Representatives, *The Solyndra Failure*, 17.
26. Ibid., 18.
27. Ibid., 21–22.
28. Ibid., 22–23.
29. Ibid., 23–24.
30. Ibid., 27.
31. Ibid., 28.
32. Ibid., 29.
33. Ibid., 20.
34. Ibid., 19.
35. Ibid., 41.
36. Ibid., 46.
37. Ibid., 32.
38. Matthew Daly, "George Kaiser, Obama Donor, Discussed Solyndra Loan with White House, Emails Show," *Huffington Post*, November 9, 2011.
39. U.S. House of Representatives, *The Solyndra Failure*, 144.
40. Ibid., 32.
41. Ibid., 33.
42. Ibid., 45.
43. Ibid., 46.
44. Josie Garthwaite, "Live: Solyndra Breaks Ground on New Plant, Details

$535M DOE Project," *Gigaom,* September 4, 2009.
45. Ibid.
46. U.S. House of Representatives, *The Solyndra Failure,* 47.
47. Matthew L. Wald, "E-mails Suggest White House Weighed a 2nd Solyndra Loan Worth Almost Half a Billion Dollars," *New York Times,* October 6, 2011.
48. Carol D. Leonnig and Joe Stephens, "Lawmakers Question Loan to Solar Company," *Washington Post,* September 14, 2011.
49. Joel Rosenblatt, "Solyndra Files Plan to Reorganize in Chapter 11 Bankruptcy," *Bloomberg,* July 27, 2012.
50. Peg Brickley, "Solyndra Outlines Its Bankruptcy Repayment Plans," *The Denver Post,* July 31, 2012.
51. Carol D. Leonnig and Joe Stephens, "Solyndra: Energy Dept. Pushed Firm to Keep Layoffs Quiet until after Midterms." *Washington Post,* November 15, 2011.

Chapter 11

1. Given our early, extended discussion of environmental religion, see Robert Nelson's book, *The New Holy Wars* (Pennsylvania State University Press for the Independent Institute, 2010).
2. T. R. Bourland and R. L. Stroup, "Rent Payments as Incentives," *Journal of Forestry,* April 1996, 18–21.
3. Lynne M. Corn, "Endangered Species: Continuing Controversy," Congressional Research Service Issue Brief IB95003, 1995.
4. Brad Plumer, "Obama May Have Left Himself Wiggle Room to Approve Keystone XL," *Washington Post,* June 25, 2013.
5. Bureau of Oceans and International Environmental and Scientific Affairs, U.S. State Department. 2014, January. *Final Supplemental Environmental Impact Statement for the Keystone XL Project: Executive Summary,* January 2014, ES-24.
6. Terry Anderson, "Stopping Keystone Ensures More Railroad Tank-Car Spills," *Property and Environment Research Center,* May 14, 2014.
7. Aldo Leopold, "Conservation Economics," *Journal of Forestry,* 32 (5) (1934), 537–544.
8. Ibid.

Appendix

1. "The Sagebrush Rebellion," *U.S. News and World Report,* December 1, 1980.
2. Bureau of Land Management, Department of the Interior, *The Federal Land Policy and Management Act (FLPMA) of 1976: How the Stage Was Set for BLM's "Organic Act."*
3. National Archives, U.S. National Archives and Records Administration, *Teaching With Documents: The Homestead Act of 1862.*
4. Ibid.
5. Bureau of Land Management, *The Federal Land Policy.*
6. Andrew Jackson, "Veto Message," December 4, 1833, *The American Presidency Project.* Retrieved from: http://www.presidency.ucsb.edu/ws/?pid=67041
7. Ibid.
8. National Archives, *Teaching with Documents.*
9. Ibid.
10. T. A. Frail and Megan Megan Gambino, "Document Deep Dive: How the Homestead Act Transformed America." *Smithsonian Magazine,* May 2012.

11. National Archives, *Teaching with Documents*.
12. Bureau of Land Management, *The Federal Land Policy*.
13. National Park Service, Theodore Roosevelt and Conservation, U.S. Department of Interior, September 5, 2013; Bernard, Joan Kelly Bernard and Whitney Barlow, "Teddy Roosevelt: American Museum of Natural History Celebrates Conservation President," *New York State Conservationist*, December 2012.
14. Bureau of Land Management, 200 Years of a Land Office Business: A GLO Timeline, U.S. Department of the Interior, March 2013.
15. Taylor Grazing Act, (43 U.S.C. §§ 315-3160)
16. Quoted in Char Miller, *Public Lands, Public Debates: A Century of Controversy* (Corvallis: Oregon University Press, 2012).
17. Nick Spitzer, "The Story of Woody Guthrie's 'This Land Is Your Land,'" *NPR Music*, February 15, 2012.
18. Public Land Law Review Commission, Bureau of Land Management. *One Third of the Nation's Land: A Report to the President and to the Congress by the Public Land Law Review Commission* (Washington DC: Government Printing Office, June 1970).
19. Federal Land Policy and Management Act (43 U.S. Code § 1701).
20. Bureau of Land Management, *The Federal Land Policy*.
21. FLPMA, (43 C.F.R. $ 3809.0-5(k) (1999))
22. American Lands Council, "To Secure and Defend Local Control of Land Access, Land Use and Land" Ownership [brochure].
23. (FLPMA of 1976 § 1701(a)(1))
24. American Lands Council, "To Secure and Defend."
25. Bureau of Land Management, *The Federal Land Policy*.
26. Pamela Manson, D. J. Summers, and Cathy Mckitrick, "Not Much Life Left for Utah Graveyards." *The Salt Lake Tribune*, May 27, 2013.
27. Ibid.
28. Matt Canham, "House Bill Would Give Fruit Heights a Cemetery," *The Salt Lake Tribune*, June 11, 2013.
29. (*Code of Federal Regulations*, 2011, Title 36, Section 254)
30. Ibid., 254.32.
31. University of Nevada, *Special Collections: A Guide to the Records of Sagebrush Rebellion: Collection No. 85-04*, June 22, 2008. Retrieved from: http://knowledgecenter.unr.edu/specoll/mss/85-04.html
32. Dennis Romboy, "Utah Poised to Lead Another Sagebrush Rebellion over Federal Land," *Deseret News*, March 4, 2012. Retrieved from: http://www.deseretnews.com/article/865551474/Utah-poised-to-lead-another-Sagebrush-Rebellion.html?pg=all
33. (FLPMA of 1976 § 1711(c)(2))
34. (FLPMA of 1976 § 1701(a)(8))
35. Ibid.
36. (FLPMA of 1976 § 1701(12))
37. Southern Utah Wilderness Alliance, "About SUWA."
38. Lewis and Clark's journals are usually used as a general baseline for Northern America.
39. *Southern Utah Wilderness Alliance v. Norton*, 2002.
40. Bureau of Land Management, *The Federal Land Policy*.
41. Southern Utah Wilderness Alliance v. Bureau of Land Management. 2006. 425 F.3d 735. Retrieved from: http://www.suwa.org/wp-content/uploads/SUWA-v-BLMopinion-Westlaw_Document_11_24_58-_2_.pdf

42. Southern Utah Wilderness Alliance, "County Official Defies Federal Agency Authority in Recapture Canyon," 2014.
43. Broadwalk – Protect Greater Canyonlands, UT. (2012). Retrieved June 5, 2015, from http://greatoldbroads.org/broadwalk-protect-greater-canyonlands-ut/
44. Mary Richards, "Group to Ask President Obama to Turn Greater Canyonlands into National Monument," *Deseret News,* November 12, 2012.
45. Southern Utah Wilderness Alliance, "Protecting Greater Canyonlands."
46. (FLPMA of 1976 § 1702(h))
47. Carmel Finley and Naomi Oreskes, "Food for Thought: Maximum Sustained Yield: A Policy Disguised as Science," *ICES Journal of Marine Science,* 70 (2) (2013): 245–250.
48. Ibid.
49. Daniel B. Botkin, "New European Fisheries Council Policy Shows That the Balance of Nature Belief Is Alive and Vigorous," May 30, 2013.
50. Christina Johnson, "The Problem is Maximum Sustainable Yield." Sea Grant California web site at http://www-csgc.ucsd.edu/NEWSROOM/NEWSRELEASES/Carmel_Finley.html. Retrieved 22 June 2015.
51. Botkin, "New European Fisheries."
52. The Wild Free-Roaming Horses and Burros Act (16 U.S.C. § 1333(a)).
53. Ibid., (b)(2).
54. Bureau of Land Management, *BLM Calico Mountains Complex Wild Horse Gather: Fact Sheet: Wildlife and Environmental Considerations.*
55. Ibid.
56. *In Defense of Animals v. Ken Salazar,* 675 F. Su2d 89 (2010), 97–98.
57. (FLPMA § 1732(b))
58. *Theodore Roosevelt Conservation Partnership v. Salazar,* 661 F.3d 66 (2011).

References

Ackerman, Bruce A. and William T. Hassler. 1981. *Clean Coal Dirty Air.* New Haven and London: Yale University Press.

Adams, Jonathan S. and Thomas O. McShane. 1992. *The Myth of Wild Africa: Conservation Without Illusion.* New York, New York: W.W. Norton.

Adams, Michael. 2005. "Beyond Yellowstone? Conservation and Indigenous rights in Australia and Sweden." University of Wollongong Research Online. Retrieved from: http://ro.uow.edu.au/cgi/viewcontent.cgi?article=1027&context=scipapers

Adler, Jonathan H. 2000. "Clean Politics, Dirty Profits." *Political Environmentalism.* Edited by Terry L. Anderson. Stanford: Hoover Institution Press.

———. 2002. "Fables of the Cuyahoga: Reconstructing a History of Environmental Protection." *Fordham Environmental Law Journal,* 14: 89–146. Retrieved from: http://law.cwru.edu/faculty/adler_jonathan/publications/fables_of_the_cuyahoga.pdf

———. 2004, September 3. "The Fable of Federal Environmental Regulation: Reconsidering the Federal Role in Environmental Protection" *Case Western Reserve Law Review,* 55: 96–101. Retrieved from: http://law.case.edu/faculty/adler_jonathan/publications/FablesFedReg.pdf

———. Winter 2008. "Anti-Conservation Incentives." *Regulation.* Retrieved from: http://object.cato.org/sites/cato.org/files/serials/files/regulation/2007/12/v30n4-6.pdf

"Agent Orange." n.d. The History Channel Website. Retrieved from: http://www.history.com/topics/agent-orange

Alfano, Peter. 2010. "NEPA at 40: Procedure or Substance." *Environmental Law Institute and The Council on Environmental Quality.* Retrieved from: http://www.eli.org/pdf/seminars/nepa/alfano.nepa.pdf

"Allen Savory, Founder and President." 2013. Savory Institute. Retrieved from: http://www.savoryinstitute.com/about-us/our-team/allan-savory/

Allen, Tony. 2013. Personal conversation between Tony Allen, formerly of Symbiotics, Inc. and Kenneth Sim regarding the City of Afton, Wyominig Micro-Hydro Project on Swift Creek.

Allin, Craig. 2002. Understanding the Wilderness Act of 1964. Cornell College. Retrieved from: http://www.cornellcollege.edu/politics/courses/allin/355/allin-wilderness%20 act.pdf

Alpert, Peter and David Western, Barry R. Noon, Brett G. Dickson, "America's Sewage System and the Price of Optimism." *Time*, August 1. 1969. Retrieved from: http://www.time.com/time/magazine/article/0,9171,901182,00.html#ixzz19KSrUirj

American Society of Civil Engineers. 2010. *Pipelines, 2010: Climbing New Peaks to Infrastructure Reliability—Renew, Rehab, and Reinvest*. Edited by George Ruchti and Tom Roode. Retrieved from: http://books.google.com/books?id=nA_QWobXZO wC&printsec=frontcover#v=onepage&q&f=false

Anderson, Terry. 2000. *Political Environmentalism: Going behind the Green Curtain*. Stanford: Hoover Institution Press.

———. 2014, May 14. "Stopping Keystone Ensures More Railroad Tank-Car Spills." *Property and Environment Research Center*. Retrieved from: http://perc.org/articles /stopping-keystone-ensures-more-railroad-tank-car-spills

Arcata Fish and Wildlife Office, U.S. Fish and Wildlife Service. 2011, July 5. *Northern Spotted Owl*.

Associated Press. 1980, October 27. "Man Ends Land Fight By Burning His House." *The Times-News*, Hendersonville, N.C. 105: (256). Retrieved from: http://news.google .com/newspapers?nid=1665&dat=19801027&id=1F8aAAAAIBAJ&sjid=xiQEAA AAIBAJ&pg=6728,5649362

———. 1995, March 13. "Report: Elk Not Overgrazing." *Missoulian*. Missoula, MT, C2.

———. 2005, June 15. "Feds May Wage War on Owls to Save Others." NBC News. Retrieved from: http://www.nbcnews.com/id/8235446/ns/us_news-environment/t /feds-may-wage-war-owls-save-others/#.Ug5N9m33NqU

Austin, J. 2007. "*Massachusetts v. Environmental Protection Agency*: Global warming, standing and the US Supreme Court." *Review Of European Community & International Environmental Law*, 16(3), 368–371. doi:10.1111/j.1467-9388.2007.00567.x

Babbitt, *Secretary of Interior v. Sweet Home Chapter of Communities for a Great Oregon*. 1995. 515 U.S. 687. Retrieved from www.lexisnexis.com/hottopics/lnacademic

Badè, William. 1916. "Introduction to 'A Thousand-Mile Walk to the Gulf' by John Muir." Retrieved from: http://www.sierraclub.org/john_muir_exhibit/writings/a_thousand _mile_walk_to_the_gulf/introduction.html

Bailey, R. 2003, December 31. "Shoot, Shovel, and Shut Up." Retrieved from: http://reason.com/archives/2003/12/31/shoot-shovel-and-shut-up

Barron, James. 2003, August 15. "Power Surge Blacks Out Northeast." *New York Times*. Retrieved from: http://www.nytimes.com/2003/08/15/nyregion/15POWE.html? ref=newyorkcityblackoutof2003&pagewanted=1

Bass, Ron. n.d. *Congressional Task Force Recommends Changes to the National Environmental Policy Act (NEPA)*. ICF International. Retrieved from: http://classwebs.spea .indiana.edu/kenricha/NEPA%20and%20the%20Forest%20Service/NEPA%2-%20

Region%206/Taskforce%20Commentaries/OregonDepartmentofTransportation_comments.pdf

Behan, Richard. December 1981. "RPA/NFMA—Time to Punt," *Journal of Forestry*, p. 802–805.

Berlau, John. 2006. *Eco-Freaks: Environmentalism is Hazardous to Your Health*. Nashville: Nelson Current.

Bernard, Joan Kelly and Whitney Barlow. 2012, December. "Teddy Roosevelt: American Museum of Natural History Celebrates Conservation President." *New York State Conservationist*. Retrieved from: http://www.dec.ny.gov/pubs/87115.html

Best, Allen. 2004, January 4. "2002 drought found to be worst in 300 years." *Summit Daily*. Retrieved from: http://www.summitdaily.com/article/20040104/NEWS/401040103

Bevington, Douglas. 2009. *The Rebirth of Environmentalism: Grassroots Activism from the Spotted Owl to the Polar Bear*. Washington: Island Press.

Biello, David. 2008, October 7. "Cylindrical solar cells give a whole new meaning to sunroof." *Scientific American*. Retrieved from: http://www.scientificamerican.com/article.cfm?id=cylindrical-solar-cells-give-new-meaning-to-sunroof

Big Cat Sticker Shack. n.d. Shoot, Shovel & Shut up (Bumper Sticker). *Amazon*. Retrieved from: http://www.amazon.com/Shoot-shovel-shut-Bumper-Sticker/dpB007BEM2GI

Blackmon, David. 2013, May 27. "The 'Sue and Settle' Racket." *Forbes*. Retrieved from: http://www.forbes.com/sites/davidblackmon/2013/05/27/the-sue-and-settle-racket/

Blankenship, Karl. 2011, July 1. *Local officials worry that TMDL actions are much too costly*. Bay Journal. Retrieved from: http://www.bayjournal.com/article/local_officials_worry_that_tmdl_actions_are_much_too_costly

Bobiec, Andrzej, Peter Landres, George Nickas. November 2004. "Managing the wild: should stewards be pilots?" *Frontiers in Ecology and the Environment*, 2: 9: 494–499. Retrieved from: http://leopold.wilderness.net/pubs/531_1.pdf

Bobkoff, Dan. 2013, August 14. "10 Years After The Blackout, How Had the Power Grid Changed?" *National Public Radio*. Retrieved from: http://www.npr.org/2013/08/14/210620446/10-years-after-the-blackout-how-has-the-power-grid-changed

Bolick, Clint. 1985, July 14. "Public Lands and Private Incentives." *The Freeman*. Retrieved from: http://www.fee.org/the_freeman/detail/public-lands-and-private-incentives#axzz2irm6RGrr

Bomey, Nathan. 2011, November 26. "A123 Systems lays off 125 workers at Michigan battery plants." *Ann Arbor News*. Retrieved from: http://www.annarbor.com/business-review/a123-systems-lays-off-125-workers-at-michigan-battery-plants/#.U1MlRrKjx4

Bonnicksen, Thomas M. 1994. "An analysis of a plan to maintain old-growth forest ecosystem." American Forest and Paper Association, Forest Resources Technical Bulletin. TB 94-3.

Botkin, Daniel B. 1990. *Discordant Harmonies: A New Ecology for the Twenty-first Century*. New York, New York: Oxford University Press.

———. 1992. *Discordant Harmonies: A New Ecology for the Twenty-first Century.* New York: Oxford University Press.

———. 2012. *The Moon in the Nautilus Shell: Discordant Harmonies Reconsidered.* New York, New York: Oxford University Press.

———. 2013, May 30. "New European Fisheries Council Policy Shows That the Balance of Nature Belief Is Alive and Vigorous." Retrieved from: http://www.danielbbotkin.com/2013/05/30/new-european-fisheries-council-policy-shows-that-the-balance-of-nature-belief-is-alive-and-vigorous/

Bourland, T. R., and R.L. Stroup. 1996, April. "Rent Payments as Incentives." *Journal of Forestry*, pp. 18–21.

Boyd, Robert. 1986. "Strategies of Indian burning in the Wilamette Valley." *Canadian Journal of Anthropology*, 5: 65–86.

Bravender, R. 2010, May 13. "EPA Issues Final 'Tailoring' Rule for Greenhouse Gas Emissions." *New York Times—Breaking News, World News & Multimedia.* Retrieved from: http://www.nytimes.com/gwire/2010/05/13/13greenwire-epa-issues-final-tailoring-rule-for-greenhouse-32021.html

Brickley, Peg. 2012, July 31. "Solyndra outlines its bankruptcy repayment plans." *The Denver Post.* Retrieved from: http://www.denverpost.com/business/ci_21195633/solyndra-outlines-its-bankruptcy-repayment-plans

Bryner, Gary C. 1987. *Bureaucratic Discretion: Law and Policy in Federal in Regulatory Agencies.* New York: Pergamon Press.

Buck, E. H., Corn, M. L., Alexander, K., Sheikh, P. A., & Meltz, R. 2012, June 14. The Endangered Species Act (ESA) in the 112th Congress: Conflicting Values and Difficult Choices. *Federation of American Scientists.* Retrieved from: www.fas.org/sgp/crs/misc/R41608.pdf

Budiansky, Stephen. 1995. *Nature's Keepers: The New Science of Nature Management.* New York, New York: Free Press.

Bureau of Land Management. n.d.(a). *The Federal Land Policy and Management Act (FLPMA) of 1976: How the Stage Was Set for BLM's "Organic Act."* Retrieved from: http://www.blm.gov/flpma/organic.htm

———. n.d.(b). *BLM Calico Mountains Complex Wild Horse Gather: Fact Sheet: Wildlife and Environmental Considerations.* Retrieved from: http://www.blm.gov/pgdata/etc/medialib/blm/nv/field_offices/winnemucca_field_office/programs/wild_horse_burro/winnemucca_wild_horse/calico_wild_horseo.Par.64861.File.dat/calico_fact_sheet.pdf

———. Department of the Interior. 2012, July 13. BLM Manual 6330—Management of BLM Wilderness Study Areas. Retrieved from: http://www.blm.gov/pgdata/etc/medialib/blm/wo/Information_Resources_Management/policy/blm_manual.Par.31915.File.dat/6330.pdf

———. 2013, March. 200 Years of a Land Office Business: A GLO Timeline. U.S. Department of the Interior. Retrieved from: http://www.blm.gov/es/st/en/prog/glo/glo_timeline.html

Bureau of Mines. 1967. Ringelmann Smoke Chart. U.S. Department of Interior. Retrieved October 29, 2013 from: http://www.cdc.gov/niosh/mining/UserFiles/works/pdfs/ic8333.pdf

Bureaucracy. Retrieved from: http://freethecommons.com/2012/06/07/nepa-as-a-case-study-of-our-broken-environmental-bureaucracy/

Bureau of Oceans and International Environmental and Scientific Affairs, U.S. State Department. 2014, January. "Chapter Three: Affected Environment." *Final Supplemental Environmental Impact Statement for the Keystone XL Project*. Retrieved from: http://keystonepipeline-xl.state.gov/documents/organization/221168.pdf

———. 2014, January. *Final Supplemental Environmental Impact Statement for the Keystone XL Project: Executive Summary*. Retrieved from: http://keystonepipeline-xl.state.gov/documents/organization/221135.pdf

CafePress. n.d. Anti Wolf Gifts & Merchandise. *CafePress*. Retrieved from: http://www.cafepress.com/+anti-wolf+gifts

California Sea Grant. Retrieved from: http://www-csgc.ucsd.edu/NEWSROOM/NEWSRELEASES/Carmel_Finley.html

Canham, Matt. 2013, June 11. "House bill would give Fruit Heights a cemetery." *The Salt Lake Tribune*. Retrieved from: http://www.sltrib.com/sltrib/politics/56445375-90/cemetery-fruit-heights-bill.html.csp

Caputo, Drew. November 1997. "A Job Half Finished: The Clean Water Act After Twenty-Five Years." *The Environmental Law Reporter*, 27: 10574. Retrieved from: http://elr.info/news-analysis/27/10574/job-half-finished-clean-water-act-after-twenty-five-years

Caro, Tim. 2010. *Conservation by Proxy: Indicator, Umbrella, Keystone, Flagship, and Other Surrogate Species*. Washington, DC: Island Press.

Carson, Rachel. 1962. *Silent Spring*. Houghton Mifflin.

———. 2002. *Silent Spring*. 1962. Reprint, Boston: Houghton Mifflin Company.

Case Western Reserve University. 1997, June 25. *The Encyclopedia of Cleveland History: Cuyahoga River Fire*. Retrieved from http://ech.cwru.edu/ech-cgi/article.pl?id=CRF1

Center for Biological Diversity. n.d.(a). Defending Endangered Species. *Center for Biological Diversity*. Retrieved from: http://www.biologicaldiversity.org/programs/biodiversity/defending_endangered_species

———. n.d.(b). 110 Success Stories for Endangered Species Day 2012. *Center for Biological Diversity*. Retrieved from: http://www.esasuccess.org/report_2012.html

Chase, Alston. 1997. "The Dark Side." *Range*. 5(2): 4–8, 62.

Chevron U.S.A., Inc. v. Natural Resources Defense Council, Inc. (Chevron v. NRDC) n.d. *Legal Information Institute*. Retrieved from: http://www.law.cornell.edu/supct/html/historics/USSC_CR_0467_0837_ZO.html

Clark, D. A., Anthony, R. G., & Andrews, L. S. 2013. Relationship between Wildfire, Salvage Logging, and Occupancy of Nesting Territories by Northern Spotted Owls. *The Journal of Wildlife Management*, 77(4), 672–688. Retrieved from the Wiley Online Library database.

Clayton, M. 2007, September 7. "New Tool to Fight Global Warming: Endangered Species Act?" *The Christian Science Monitor.* Retrieved from: http://www.csmonitor.com/2007/0907/p03s03-usgn.html

Coates, Dennis and Brad R. Humphreys. 2001. "The Economic Consequences of Professional Sports Lockouts and Strikes," *Southern Economic Journal*, Vol. 67, No. 3 (January 2001), pp. 737–747.

Code of Federal Regulations (CFR). 1978. U.S. Government Printing Office. Title 40. Council on Environmental Quality. Retrieved from: http://www.gpo.gov/fdsys/search/pagedetails.action?collectionCode=CFR&searchPath=Title+40%2FChapter+V%2FPart+1502&granuleId=CFR-2012-title40-vol34-part1501&packageId=CFR-2012-title40-vol34&oldPath=Title+40%2FChapter+V%2FPart+1502&fromPageDetails=true&collapse=false&ycord=617

———. 2011. US Government Printing Office. Title 36. Landownership Adjustments. Retrieved from: http://www.gpo.gov/fdsys/granule/CFR-2011-title36-vol2/CFR-2011-title36-vol2-part254/content-detail.html

Codevilla, Angelo M. 2009, April. "Scientific Pretense vs. Democracy." *The American Spectator.* Retrieved on October 28, 2013 from: http://spectator.org/archives/2009/04/14/scientific-pretense-vs-democra/print

Cohen, Bonner. 2009, July 13. "Red Rock Wilderness Bill." Committee For A Constructive Society. Retrieved from: http://www.cfact.org/2009/07/13/red-rock-wilderness-bill/

Cole, David N. January 2001. "Management Dilemmas That Will Shape Wilderness in the 21st Century." *Journal of Forestry*, 99(1): 4–8. Retrieved from: http://www.fs.fed.us/rm/pubs_other/rmrs_2001_cole_d001.pdf

Committee on Environment and Public Works. US Senate. 2010, December 10. *Senate Report 111-361: Clean Water Restoration Act.* Retrieved from: http://www.gpo.gov/fdsys/pkg/CRPT-111srpt361/html/CRPT-111srpt361.htm

Committee on Natural Resources. U.S. House of Representatives. 2013, February 7. *Subcommittee Chairman Bishop Leads Request for Comprehensive Analysis of NEPA's Costs and Regulatory Burdens on Taxpayers.* Retrieved from: http://naturalresources.house.gov/news/documentsingle.aspx?DocumentID=319306

Commoner, Barry. 1971. *The Closing Circle: Nature, Man, and Technology.* Knopf.

Congressional Budget Office. U.S. Congress. 2012, June 20. H.R. 4965, a bill to preserve existing rights and responsibilities with respect to waters of the United States, and for other purposes. Retrieved from: http://www.cbo.gov/publication/43344

Congressional Research Service. 2006, March 8. Energy Policy Act of 2005: Summary and Analysis of Enacted Provisions. Retrieved from: http://www.circleofblue.org/waternews/wp-content/uploads/2010/08/CRS-Summary-of-Energy-Policy-Act-of-2005.pdf

Connolly, Kim Diana. 2006. *US Supreme Court Rapanos and Carabell Wetlands Cases.* University of South Carolina School of Law. Retrieved from: http://law.sc.edu/wetlands/rapanos-carabell/rapanos.shtml

Connor, Steve. 1994. The Independent. *Ice pack reveals Romans' air pollution.* Retrieved from: http://www.independent.co.uk/news/uk/ice-pack-reveals-romans-air-pollution-1450572.html

Constitutional Defense Council. 2012, November 14. Report on Utah's Transfer of Public Lands Act H.B. 148.

Cook, Noble D., & W. G. Lovell (Eds.). 1992. *Secret Judgments of God: Old World Disease in Colonial Spanish America.* Norman, OK: University of Oklahoma Press.

Copeland, Claudia. 2010, April 23. *Clean Water Act: A Summary of the Law.* Prepared for Congressional Research Service. Retrieved from: http://www.oakparkusd.org/cms/lib5/CA01000794/Centricity/Domain/338/clean%20water%20act%20summary.pdf

———. 2012, September 21. Clean Water Act and Pollutant Total Maximum Daily Loads (TMDLs). Prepared for Congressional Research Service. Retrieved from: http://www.fas.org/sgp/crs/misc/R42752.pdf

Corn, Lynne M. 1995. *Endangered Species: Continuing Controversy. Congressional Research.* Service Issue Brief IB95003.

Cornell Lab of Ornithology. n.d. Barred Owl. *Cornell Lab of Ornithology.* Retrieved from: http://www.allaboutbirds.org/guide/barred_owl/id

Cornell University Law School. n.d.(a). 16 U.S.C. §—Title 16. Retrieved from: http://www.law.cornell.edu/uscode/text/16/460l-4

Cornell University Law School. n.d.(b). *Babbitt v. Sweet Home Chapt. Comms. for Ore.,* 515 U.S. 687 (1995). Retrieved from: http://www.law.cornell.edu/supct/html/94-859.ZO.html

Corwin, Miles. 1989, January 28. "The Oil Spill Heard 'Round the Country!" *Los Angeles Times.* Retrieved from: http://www2.bren.ucsb.edu/~dhardy/1969_Santa_Barbara_Oil_Spill/Home.html

Coulter, Michael. 2005, August 1. *House Task Force Hears Testimony on Improving Decades-Old Environmental Law.* The Heartland Institute. Retrieved from: http://news.heartland.org/newspaper-article/2005/08/01/house-task-force-hears-testimony-improving-decades-old-environmental-la

Council on Environmental Quality (CEQ). 2003. NEPA Task Force. *Modernizing NEPA Implementation.* U.S. Executive Office of the President. p. 121.

Cox, Tony. August 2011. "Reassessing the Human Health Benefits from Cleaner Air." George Mason University. Retrieved on October 28, 2013 from: http://www.cmpa.com/pdf/ReassessingCleanAirAug22.pdf

Crankcase emission control. n.d. Retrieved from: http://www.cdxetextbook.com/fuelSys/emission/control/crankcase.html

Cronon, William. 1995. *Uncommon Ground: Toward Reinventing Nature.* New York: W. W. Norton & Company.

———. 1996. *Uncommon Ground: Rethinking the Human Place in Nature.* New York: W. W. Norton & Company.

———. n.d. *The Trouble with Wilderness.* Retrieved from: http://www.williamcronon.net/writing/Trouble_with_Wilderness_Main.html

Cruz, Gilbert. 2010, May 3. "Top 10 Environmental Disasters." *Time*. Retrieved from: http://content.time.com/time/specials/packages/article/0,28804,1986457_1986501_1986441,00.html

"Cuyahoga River Fire." n.d. Ohio History Central. Retrieved from: http://www.ohiohistorycentral.org/w/Cuyahoga_River_Fire?rec=1642

Dahl, Robert A. 1956. *A Preface to Democratic Theory*. University of Chicago Press.

Daly, Matthew. 2011, November 9. "George Kaiser, Obama donor, discussed Solyndra Loan with White House, Emails Show." *Huffington Post*. Retrieved from: http://www.huffingtonpost.com/2011/11/09/george-kaiser-solyndra_n_1084568.html

Database of State Incentives for Renewables & Efficiency. 2011. Federal Incentives/Policies for Renewables & Efficiency. Retrieved from: http://www.dsireusa.org/incentives/allsummaries.cfm?State=us&re=1&ee=1

Daubenmire, R. 1985. "The Western Limits of the Range of American Bison." *Ecology*. 66:622–624.

Defenders of Wildlife. n.d.(a). Northern Spotted Owl. *Defenders of Wildlife*. Retrieved from: http://www.defenders.org/northern-spotted-owl/basic-facts

———. n.d.(b). Endangered Species Act 101. *Defenders of Wildlife*. Retrieved from: http://www.defenders.org/endangered-species-act/endangered-species-act?gclid=CIncuu-gj7kCFZBaMgodIXUArwdenied-study-potential-economic-impact-permitting-challenges-facing-proposed-energy

de Steiguer, J. E. 2006. *The Origins of Modern Environmental Thought*. Tuscon: University of Arizona Press.

Dewey, T., & Smith, J. 2002. Canis Lupus. *Animal Diversity Web*. Retrieved from: http://animaldiversity.ummz.umich.edu/accounts/Canis_lupus/

Diamond, Jared. 1988. "The Golden Age that Never Was." *Discover*. 9(12): 70–79.

Dicks, N. 2004, May 20. Statement of Congressman Norm Dicks re Hatchery Salmon. *Project Vote Smart*. Retrieved from: http://votesmart.org/public-statement/41163/#.UcsEQJyWb3U

DiSavino, Scott. 2013, August 11. "RPT-Ten years after NE blackout, U.S. power grid smarter, sturdier." Reuters. Retrieved from: http://www.reuters.com/article/2013/08/12/blackout-anniversary-idUSL2N0GC06R20130812

Dobyns, Henry F. 1983. *Their Numbers Become Thinned: Native American Population Dynamics in Eastern North America*. Knoxville, TN: University of Tennessee Press.

Dolan v. City of Tigard. 1994. 512 U.S. 374. Retrieved from: http://www.lexisnexis.com.dist.lib.usu.edu/hottopics/lnacademic/?

Doogan, Sean. 2013, October 10. *Both state, feds probe August EPA task force raids near Chicken*. Alaska Dispatch. Retrieved from: http://www.alaskadispatch.com/article/20131010/both-state-feds-probe-august-epa-task-force-raids-near-chicken

Dowie, Mark. 1996. *Losing Ground: American Environmentalism at the Close of the Twentieth Century*. Cambridge: The MIT Press.

―――. 2009, November 25. "Conservation: Indigenous People's Enemy No. 1?" *Mother Jones*. Retrieved from: http://www.motherjones.com/environment/2009/11/conservation-indigenous-peoples-enemy-no-1

Dubner, S. J., & Levitt, S. D. 2008, January 20. "Unintended Consequences." Freakonomics (blog), *New York Times*. Retrieved from: http://www.nytimes.com/2008/01/20/magazine/20wwln-freak-t.html?pagewanted=print&_r=0

DuHamel, Jonathan. 2013, May 6. *How NEPA crushes productivity*. Tucson Citizen.com. Retrieved from: http://tucsoncitizen.com/wryheat/2013/05/06/how-nepa-crushes-productivity/

Duhigg, Charles. 2009, September 12. "Clean Water Laws are Neglected, at a Cost in Suffering." *New York Times*. Retrieved October 22, 2013 from: http://www.nytimes.com/2009/09/13/us/13water.html?pagewanted=all&_r=0

Duhigg, Charles and Janet Roberts. 2010, February 28. "Rulings Restrict Clean Water Act, Foiling E.P.A." *New York Times*. Retrieved from: http://www.nytimes.com/2010/03/01/us/01water.html?hp

Durban, Dick. n.d. "Durbin and Hitchey Introduce Bill to Protect America's Red Rock Wilderness." Retrieved from: http://www.durbin.senate.gov/public/index.cfm/pressreleases?ID=7a3d0d17-f543-4a80-b6d8-4ae9f7cd60ab

Dwyer, L. E., Murphy, D. D., and Ehrlich, P. R. 1995. Property Rights Case Law and the Challenge to the Endangered Species Act. *Conservation Biology*, 9(4), 725–741. Retrieved from: http://www.life.illinois.edu/ib/451/Dwyer%20(1995).pdf

"Ecological Balance." n.d. WWF Global. Retrieved from: http://wwf.panda.org/about_our_earth/teacher_resources/webfieldtrips/ecological_balance/

Economic Realities in the Yellowstone Region. 2012, July 2. Bozeman, MT: The Wilderness Society.

Eco-Now. 2012, July 2. "Endangered Species Act Success Stories." Retrieved from: http://econowblog.blogspot.com/2012/07/endangered-species-act-success-stories.html

Ehrlich, Paul R. 1968. *The Population Bomb*. New York: Ballantine Books.

Emerson, Ralph Waldo. 1894. *Nature: Addresses/Lectures*. Boston and Cambridge: James Munroe and Company: Cambridge Massachusetts.

eNature. n.d. FieldGuides. *eNature: America's Wildlife Resource*. Retrieved from: http://www.enature.com/fieldguides/detail.asp?recNum=FI0395

Endangered Species Act. 1973. U.S. Code. Title 16. Sections 1531–1544.

"Endangered Species Act." n.d. *ThinkQuest: Library*. Retrieved from: http://library.thinkquest.org/26026/Politics/endangered_species_act.html

Energy Policy Act of 2005. U.S. Statutes at Large. 2005. Pub.L. 109–58.

Engelberg, D., K. Butler, C. Brunton, A. Chirigotis, J. Hamann, and M. Reed. 2011, December 9. *EPA Must Improve Oversight of State Enforcement*. Prepared for U.S. Environmental Protection Agency, Office of Inspector General. Retrieved from: http://www.epa.gov/oig/reports/2012/20111209-12-P-0113.pdf

Enviro Defenders. n.d. *Legal Handbook for Environmental Activists: NEPA & CEQA*. Retrieved from: http://envirodefenders.org/legal/nepa.html

Environmental Protection Agency. 1992. "The Guardian: Origins of the EPA". Retrieved from: http://www2.epa.gov/aboutepa/guardian-origins-epa

———. 2009(a), March 12. *National Pollutant Discharge Elimination System (NPDES)*. Retrieved from: http://cfpub.epa.gov/npdes/

———. 2009(b). Endangerment and Cause or Contribute Findings for Greenhouse Gases under Section 202(a) of the Clean Air Act. Retrieved October 26, 2013 from: http://www.epa.gov/climatechange/endangerment/

———. 2010(a). Prevention of Significant Deterioration and Title V Greenhouse Gas Tailoring Rule. *US Environmental Protection Agency*. Retrieved from: http://www.epa.gov/apti/video/TailoringRule/tailoring.pdf

———. 2010(b), December 29. *Chesapeake Bay TMDL Executive Summary*. Retrieved from: http://www.epa.gov/reg3wapd/pdf/pdf_chesbay/FinalBayTMDL/BayTMDLExecutiveSummaryFINAL122910_final.pdf

———. Region 6 Office. 2011, October 4. *Clean Water Act*. Retrieved from: http://www.epa.gov/region6/6en/w/cwa.htm (From Chapter 9: Clean Water Act)

———. 2012, August 27. *What is Nonpoint Source Pollution?* Retrieved from: http://water.epa.gov/polwaste/nps/whatis.cfm

———. 2013(a). Vehicle Standards and Regulations. Retrieved October 21, 2013 from: http://www.epa.gov/otaq/standards.htm

———. 2013(b). "Overview of EPA Authorities for Natural Resource Managers Developing Aquatic Invasive Species Rapid Response and Management Plans: CWA Section 404—Permits to Discharge Dredged or Fill Material." Retrieved from: http://water.epa.gov/type/oceb/habitat/cwa404.cfm (From Chapter 3: Federal Lands Policy).

———. 2013(c). "Toxic Substances Control Act." Retrieved from: http://www.epa.gov/agriculture/lsca.html#Summary of Toxics Substances Control Act (TSCA) (From Chapter 2: History of Environmental Movement)

———. 2013(d). History of the Clean Air Act. Retrieved from: http://www.epa.gov/air/caa/amendments.html (From Chapter 8: Clean Air Act)

———. 2013(e). Overview of EPA Authorities for Natural Resource Managers Developing Aquatic Invasive Species Rapid Response and Management Plans: CWA Section 404—Permits to Discharge Dredged or Fill Material. Retrieved from: http://water.epa.gov/type/oceb/habitat/cwa404.cfm (From Chapter 9: Clean Water Act)

———. 2013(f), July 25. *Priority Pollutants*. Retrieved from: http://water.epa.gov/scitech/methods/cwa/pollutants.cfm (From Chapter 9: Clean Water Act)

———. 2013(g), March 12. *Clean Water Act (CWA)*. Retrieved from: http://www.epa.gov/agriculture/lcwa.html (From Chapter 9: Clean Water Act)

———. n.d. The Clean Air Act in a Nutshell: How It Works. Retrieved from: http://www.epa.gov/air/caa/pdfs/CAA_Nutshell.pdf (From Chapter 8: Clean Air Act)

Federal Land Policy and Management Act. 1976. U.S. Code. Title 43. Section 1701 et seq.

Ferguson, E.B. 2013, September 24. *Farm lobby likely to appeal Chesapeake Bay pollution diet ruling*. Capital Gazette. Retrieved from: http://www.capitalgazette.com/news/environment/farm-lobby-likely-to-appeal-chesapeake-bay-pollution-diet-ruling/article_09a83b93-e2b7-577f-aa2c-ea8c76fc5d3b.html

Finley, Carmel and Naomi Oreskes. 2013. "Food for Thought: Maximum sustained yield: a policy disguised as science." *ICES Journal of Marine Science*, 70: (2): 245–250. Retrieved from: http://icesjms.oxfordjournals.org/content/70/2/245.full.pdf+html

Fischer, G. 1948. Award Ceremony Speech. The Nobel Foundation. Retrieved from: http://www.nobelprize.org/nobel_prizes/medicine/laureates/1948/press.html

Fish and Wildlife Service. 2013, January 10. *Digest of Federal Resource Laws*. Retrieved from: http://www.fws.gov/laws/lawsdigest/Resourcelaws.html

Flatt, Victor B. 1997, September 1. "Dirty River Runs Through It (The Failure of Enforcement in the Clean Water Act)." *Boston College Environmental Affairs Law Review*, 25: 1–45. Retrieved from: http://lawdigitalcommons.bc.edu/cgi/viewcontent.cgi?article=1269&context=ealr&seiredir=1&referer=http%3A%2F%2Fscholar.google.com%2Fscholar%3Fq%3Dclean%2Bwater%2Bact%2Bflatt%26btnG%3D%26hl%3Den%26as_sdt%3D0%252C45#search=%22clean%20water%20act%20oflatt%22

Fong, J. 2011, April 29. "Fox Sees Conspiracy in Effort to Protect Lizard Species." *Media Matters*. Retrieved from: mediamatters.org/blog/2011/04/29/fox-sees-conspiracy-in-effort-to-protect-lizard/179200

Foreman, Dave. 1985. *Ecodefense: A Field Guide to Monkeywrenching*. Tucson, AZ: Earth First! Books.

Frail, T.A. and Megan Gambino. 2012, May. "Document Deep Dive: How the Homestead Act Transformed America." *Smithsonian Magazine*. Retrieved from: http://www.smithsonianmag.com/history-archaeology/How-the-Homestead-Act-Transformed-America.html

Free the Commons. 2012, June 7. "NEPA as a Case Study of our Broken Environmental Bureaucracy." Retrieved from: http://freethecommons.com/2012/06/07/nepa-as-a-case-study-of-our-broken-environmental-bureaucracy/

Fristik, Richard. 2012, November. *Mitigation Under NEPA: Failed Promises?* Prepared for U.S. Department of Agriculture. Retrieved from: http://dukespace.lib.duke.edu/dspace/bitstream/handle/10161/5980/R.%20Fristik_Duke%20Capstone%20Paper_revised_10-31-12.pdf?sequence=1

Frome, Michael. 2007. *Rebel on the Road: And Why I Was Never Neutral*. Truman State University Press: Kirksville, Missouri. Retrieved from: http://www.amazon.com/Rebel-Road-And-Never-Neutral/dp/1931112657

Galilei, Galileo. 1615. Letter to the Grand Duchess Christina of Tuscany. Retrieved from: http://www.fordham.edu/halsall/mod/galileo-tuscany.asp

Garthwaite, Josie. 2009, September 4. Live: Solyndra breaks ground on new plant,details $535M DOE project. *Gigaom*. Retrieved from: http://gigaom.com/cleantech/live-solyndra-breaks-ground-on-new-plant-details-535m-doe-project/

Geist, Valerius. 1996. *Buffalo Nation: History and Legend of the North American Bison*. McGregor, MN: Voyageur Press.

General Accounting Office. 1972, March 23. *Water Pollution Abatement Program: Assessment of Federal And State Enforcement Efforts.* Retrieved from: http://www.gao.gov/assets/210/200061.pdf

Gerhart, M. 2009. "Climate Change and the Endangered Species Act." *Ecology Law Quarterly*, 36: (167).

Gies, E. 2012, May 30. "Report: Endangered Species Act Works.' *Forbes.* Retrieved from: http://www.forbes.com/sites/ericagies/2012/05/30/endangered-species-act-works-says-report/

Goklany, Indur. 1999. *Clearing the Air.* Washington, D.C.: The Cato Institute.

Goodell, Jeff. 2013, July 7. "Meet America's Most Creative Climate Criminal." *Rolling Stone Politics.* Retrieved from: http://www.rollingstone.com/politics/blogs/national-affairs/meet-america-s-most-creative-climate-criminal-20110707

Gore, Al. 1992. *Earth in the Balance: Ecology and the Human Spirit.* Boston: Houghton Mifflin Company.

———. 2006. *Earth in the Balance: Ecology and the Human Spirit.* Rodale.

Gore, D'Angelo, and Eugene Kiely. 2011, October 7. Obama's Solyndra problem. *FactCheck.org.* Retrieved from: http://www.factcheck.org/2011/10/obamas-solyndra-problem/

Gottesfeld, L.M.J. 1994. "Aboriginal burning for vegetative management in northeast British Columbia." *Human Ecology*, 22: 171–188.

Government Accountability Office (GAO). 1994, December 20. *Endangered Species Act: Information on Species Protection on Nonfederal Lands.* Retrieved from: http://www.gao.gov/products/RCED-95-16

———. 2009, October 15. *Clean Water Act: Longstanding Issues Impact EPA's and States' Enforcement Efforts.* Retrieved from: http://www.gao.gov/new.items/d10165t.pdf

Govtrack.us. 2013. *H.R. 933: Fruit Heights Land Conveyance Act.* Retrieved from: http://www.govtrack.us/congress/bills/113/hr993

Grayson, Donald K. 1993. *The Desert's Past: A Natural Prehistory of the Great Basin.* Washington, D.C.: Smithsonian Institute Press.

"Green-energy ideas so crazy they just might work." n.d. NBCNEWS.com. Retrieved from: http://www.msnbc.msn.com/id/38730065/ns/technology_and_science-future_of_energy/t/green-energy-ideas-so-crazy-they-just-might-work/#.UBxKj7R8BK1

Greenhouse, Linda. 2005, October 12. "Supreme Court Takes Up 2 Cases Challenging Powers of U.S. Regulators to Protect Wetlands." *New York Times.* Retrieved from: http://www.nytimes.com/2005/10/12/politics/12scotus.html?pagewanted=all

Greenstone, Michael. December 2002. "The Impacts of Environmental Regulations on Industrial Activity: Evidence from the 1970 and 1977 Clean Air Act Amendments and the Census of Manufactures." *Journal of Political Economy*, 110: 1175, 1219. Retrieved on October 29, 2013 from: http://www.jstor.org/stable/10.1086/342808

Grumbine, R. E. March 1994. "What is Ecosystem Management?" *Conservation Biology.* 8(1): 27–34.

Harvey, Mark W. T. 2005. *Wilderness Forever: Howard Zahniser and the Path to the Wilderness Act.* Seattle: University of Washington Press.

Hastings, D. 2011, December 6. Excessive Endangered Species Act Litigation Threatens Species Recovery, Job Creation and Economic Growth. *House Committee on Natural Resources*. Retrieved from: http://naturalresources.house.gov/news/documentsingle.aspx?DocumentID=271408

———. 2012, May 18. "Hastings: Time to Improve the Endangered Species Act." *Washington Times*. Retrieved from: http://www.washingtontimes.com/news/2012/may/18/time-to-improve-the-endangered-species-act/

Haws, J. Matthew. 2012. "Analysis Paralysis: Rethinking the Court's Role in Evaluating EIS Reasonable Alternatives." *University of Illinois Law Review*, 2012: 537–576.

Hayek, F.A. (1948). "The Use of Knowledge in Society." *Individualism and Economic Order*, pp. 77–91.

Henderson, Nia-Malika. 2012, May 31. "Mitt Romney visits Solyndra headquarters, knocks President Obama." *Washington Post*. Retrieved from: http://articles.washingtonpost.com/2012-05-31/politics/35455861_1_solyndra-failure-mitt-romney-free-enterprise

Hill, Peter, and Roger E. Meiners. 1998. *Who Owns The Environment?* Lanham, Maryland: Rowman & Littlefield Publishers, Inc.

Hines, William N. 1966. "Nor Any Drop to Drink: Public Regulation of Water Quality Part I: State Pollution Control Programs." *Iowa Law Review*, 52: 186–235. Retrieved from: http://heinonline.org/HOL/Page?collection=journals&handle=hein.journals/ilr52&div=18&id=&page=#214

"The History of the Endangered Species Act." n.d. *The Thoreau Institute*. Retrieved from: http://www.ti.org/ESAHistory.html

Hoffinger, Fran. 1982. "Environmental Impact Statements: Instruments for Environmental Protection or Endless Litigation?" *Fordham Urban Law Journal*, 11: 527–566.

Hopper, R. 2012, June 5. "Inflated Endangered Species Act 'Success Stories' Revealed." *PLF Liberty Blog*. Retrieved from: http://blog.pacificlegal.org/2012/inflated-endangered-species-act-success-stories-revealed/

Hopson, Mark Christopher. 2011. "The Wilderness Myth: How The Failure of the American National Park Model Threatens the Survival of the Iyaelima Trive and the Bonobo Chimpanzee." *Environmental and Earth Law Journal*, 1:1 Retrieved from: http://lawpublications.barry.edu/cgi/viewcontent.cgi?article=1003&context=ejejj

Hornaday, William T. 1913. *Our Vanishing Wildlife*. New York: New York Zoological Society.

House Committee on Natural Resources. 2005, November 17. *NEPA: Lessons Learned and Next Steps*. House of Representatives Hearing, November 17, 2005, U.S. Government Printing Office.

———. 2009, October 1. "Utah Congressional Delegation Expresses Bipartisan Opposition to America's Red Rock Wilderness Act." Retrieved from: http://naturalresources.house.gov/news/documentsingle.aspx?DocumentID=147628

———. n.d. Endangered Species Act. *House Committee on Natural Resources*. Retrieved from: http://naturalresources.house.gov/esa/

Hrab, Neil. 2004, March. "Greenpeace, Earth First!, PETA: Radical Fringe Tactics Move Toward Center Stage." Organizational Trends. Capital Research Center. Retrieved from: http://capitalresearch.org/pubs/pdf/03_04_OT.pdf

Hudson, Audrey. 2012, September 13. "Environmentalists Oppose Obama Plan to Develop Solar Energy." *Human Events.*

In Defense of Animals v. Ken Salazar. 2010. 675 F. Supp. 2d 89, 97–98. Retrieved from: http://scholar.google.com/scholar_case?case=15989834399269600367&hl=en&as_sdt=2&as_vis=1&oi=scholarr

International Center for Technology Assessment. October 1999. "Petition for Rulemaking and Collateral Relief Seeking the Regulation of Greenhouse Gas Emissions from New Motor Vehicles Under?202 of the Clean Air Act." Retrieved on October 27, 2013 from: http://www.ciel.org/Publications/greenhouse_petition_EPA.pdf

Ioannidis, John P. A. 2005, August 30. "Why Most Published Research Findings Are False." *PLOS Medicine.* 2(8): 0696–0701.

Ives, Ronald L. 1942. "The beaver-mead ow complex. *Journal of Geomorphology,* 5: 191–203.

Ivester, David M. and Christian L. Marsh, 2010. "Renewable energy: Streamlining review under NEPA and the ESA." *Trends.* Retrieved from http://briscoelaw.net/wp-content/uploads/2012/01/Trends_Newsletter.pdf

Jackson, Andrew. 1833, December 4. "Veto Message." *The American Presidency Project.* Retrieved from: http://www.presidency.ucsb.edu/ws/?pid=67041

Jacobs, Lynn. 1991. *Waste of the West: Public Lands Ranching.* Privately published in Tucson, AZ: Lynn Jacobs.

Johnson, Christina. 2007, December 21. "The Problem is Maximum Sustainable Yield."

Johnson, David B. 1991. *Public Choice: An Introduction to the New Political Economy.* Mayfield Publishing Co.

Johnson, M. A. 2013, July 23. "Feds Move Ahead with plans to kill barred owls—to save spotted owls." *NBC News.* Retrieved from: http://usnews.nbcnews.com/news/2013/07/23/19645594-feds-move-ahead-with-plans-to-kill-barred-owls-to-save-spotted-owls?lite

Johnson, Newkirk L. 2005. Wilderness Forever. Retrieved from: http://www.pawild.org/articles/WildForeverRevNAJ.pdf

Kabat, Geoffrey. October 2013. "Over Reaching California EPA Regulators Promoted a False Breast Cancer Link." *Forbes Magazine.* Retrieved October 28, 2013 from: http://www.forbes.com/sites/geoffreykabat/2013/10/25/over-reaching-california-epa-regulators-promote-a-false-breast-cancer-link/

Kahneman, Daniel. 2011. *Thinking, Fast and Slow.* New York: Farrar, Strauss, and Giroux.

Kalen, Sam. 2009. "The Devolution of NEPA: How the APA Transformed the Nation's Environmental Policy." *William & Mary Environmental Law and Policy Review.* 33: 483–548. Retrieved from: http://scholarship.law.wm.edu/cgi/viewcontent.cgi?article=1035&context=wmelpr

Kareiva, Peter, Michelle Marvier, and Robert Lalasz. 2012. Conservation in the Anthropocene: Beyond Solitude and Fragility. Retrieved from: http://thebreakthrough.org/index.php/journal/past-issues/issue-2/conservation-in-the-anthropocene/

Katz, Diane & Craig Manson. 2012. "The National Environmental Policy Act." *The Heritage Foundation: Leadership for America*. Retrieved from: http://thf_media.s3.amazonaws.com/2012/EnvironmentalConservation/Chapter5-The-National-Environmental-Policy-Act.pdf

Kaufman, Leslie. 2009, February 4. "Drilling Leases Scrapped in Utah." *New York Times*. Retrieved from: http://www.nytimes.com/2009/02/05/us/05leases.html?_r=0

Kay, Charles E. 1990. "Yellowstone's Northern Elk Herd: A Critical Evaluation of the 'Natural Regulation' Paradigm." Ph.D. Dissertation, Utah State University, Logan, Utah.

———. 1994. "Aboriginal Overkill: The Role of Native Americans in Structuring Western Ecosystems." *Human Nature*. 5: 359–398.

———. 1995. Aboriginal overkill and native burning: Implications for modern ecosystem management. *Western Journal of Applied Forestry*, 10, 121–126.

———. 1996(a). "Ecosystems Then and Now: A Historical Approach to Ecosystem Management." Wilms, W.D. and J.F. Dormarr (Eds.) p. 87–89, "Proceedings of the Fourth Prairie Museum of Alberta Natural History," Occasional Paper No. 23.

———. 1996(b). *Wolf Recovery, Political Ecology, and Endangered Species*. Independent Report. Oakland, CA: Independent Institute.

———. 1997(a). "Aboriginal Arguments." *Journal of Forestry*. 95: 8.

———. 1997(b). "Aboriginal Overkill and the Biography of Moose in Western North America." *Alces*. 33: 141–164.

———. 1997(c). "Viewpoint: Ungulate Herbivory, Willows, and Political Ecology in Yellowstone." *Journal of Range Management* 50: 139–145.

———. 1997(d). "Yellowstone: Ecological Malpractice." *PERC Reports*. 15(2).

———. 2007(a). "Are lightning fires unnatural? A comparison of aboriginal and lightning ignition rates in the United States." *Tall Timbers Fire Ecology Conference*, 23:16–28. Retrieved from: http://www.idahoforwildlife.com/Website%20articles/Website%20articles/Charles%20Kay/71-Are%20lightning%20fires%20unnatural-A%20comparison%20of%20aboriginal%20and%20lighting%20fire%20ignition%20rates%20in%20the%20USA.pdf

———. 2007(b). "Were native people keystone predators? A continuous-time analysis of wildlife observations made by Lewis & Clark in 1804–1806." *Canadian Field-Naturalist*, 121: 1–16. Retrieved from: http://rliv.com/pic/LewisClark.pdf

———. 2013, July. Strata Lecture on Aboriginal Influences and the Original State of Nature. [JL1]

Kay, Charles E., and C. W. White. 1995. "Long-Term Ecosystem States and Processes in the Central Canadian Rockies: A New Perspective on Ecological Integrity and Ecosystem Management." In R. M. Linn (Ed.), *Sustainable Society and Protected Areas*. pp. 119–132. Hancock, MI: The George Wright Society.

Kay, Charles E., B. Patton, and C. White. 1994. Assessment of Long-Term Terrestrial Ecosystem States and Processes in Banff National Park and the Central Canadian Rockies. Banff, AB: Resource Conservation, Parks Canada, Banff National Park.

Keigley, R.B. 1997(a). "An Increase in Herbivory of Cottonwood in Yellowstone National Park." *Northwest Science*, 71: 127–136.

———. 1997(b). Statement of Richard Keigley to the Subcommittee on National Parks and Public Lands. Oversight Hearing on Science and Resource Management in the National Park System. Retrieved from the World Wide Web at: http://www.house.gov/resources/105cong/parks/feb27.97/keigley.html.

Knickerbocker, Brad. 1995, May 25. *Clean Water Responsibilities Slipping From Feds to States*. Christian Science Monitor. Retrieved from: http://www.csmonitor.com/1995/0525/25051.html

———. 2002, November 7. "Environmental 'Magna Carta' law under fire." *Christian Science Monitor*. Retrieved from: http://www.csmonitor.com/2002/1107/p02s02-usgn.html

Kricher, John. 2009. *The Balance of Nature: Ecology's Enduring Myth*. Princeton University Press.

Laitos, Jan G., and Rachael B. Gamble. 2008. *The Problem with Wilderness*. Harvard Environmental Law Review, 32: 503–569. Retrieved from: http://www.law.harvard.edu/students/orgs/elr/vol32_2/Laitos%20Final%20Final.pdf

LaMonica, Martin. 2009, July 16. From Onion Juice to Factory Juice. *Cnet News*. Retrieved from: http://web.archive.org/web/20090804235621/http:/news.cnet.com/8301-11128_3-10288652-54.html

Landres, Peter B., and Mark W. Brunson, Linda Merigliano, Charisse Sydoriak, Steve Mortol. 2000. "Naturalness and Wildness: The Dilemma and Irony of Managing Wilderness." USDA Forest Service Proceedings, RMPS – 15, Vol. 5: p. 377–381, p. 379. Retrieved from: http://www.fs.fed.us/rm/pubs/rmrs_p015_5.pdf

Lazerwitz, D., Bostick, M., Braun, F., and +Martel LLP. 2010. "NEPA Processes for energy projects: Unique Challenges and New Directions." *RMMLF Special Institute on NEPA*. Retrieved from: http://www.fbm.com/files/Publication/9811c207-995f-415d-b028-3b2ff0bdc5e8/Presentation/PublicationAttachment/d5b6d5d9-e600-4348-b632-3e7b02d9e26c/e66c4e73-7b75-4d50-bf56-d0c713986d99_document.pdf

Lean, Geoffery. 2012. *The Great Smog of London: the air was thick with apathy*. The Telegraph. Retrieved from: http://www.telegraph.co.uk/earth/countryside/9727128/The-Great-Smog-of-London-the-air-was-thick-with-apathy.html

Learn, S. 2011, February 5. "Northern spotted owl marks 20 years on endangered species list." *OregonLive.com*. Retrieved from: http://www.oregonlive.com/environment/index.ssf/2010/06/northern_spotted_owl_marks_20.html

Legal Ruralism. 2011, November 10. Shoot, Shovel, and Shut up. *Legal Ruralism*. Retrieved from: http://legalruralism.blogspot.com/2011/11/shoot-shovel-and-shut-up.html

Leonnig, Carol D. & Joe Stephens. 2011(a), September 14. Lawmakers question loan to solar company. *Washington Post.* Retrieved from: http://www.washingtonpost.com/politics/lawmakers-why-did-administration-risk-another-67-million-on-a-company-near-collapse/2011/09/14/gIQAD024RK_story.html

———. 2011(b), November 15. Solyndra: Energy Dept. pushed firm to keep layoffs quiet until after midterms. *Washington Post.* Retrieved from: http://articles.washingtonpost.com/2011-11-15/politics/35282189_1_solyndra-e-mails-energy-department

Leopold, Aldo. 1925. "Wilderness as a Form of Land Use." *Journal of Land and Public Utility Economics,* 1(4): 398–404.

"The Leopold Legacy: A Sand County Almanac." n.d. The Aldo Leopold Foundation. Retrieved from: http://www.aldoleopold.org/AldoLeopold/almanac.shtml

Levack, Brian, Edward Muir, Michael Mass, and Meredith Veldman. 2010. "Chapter 20: The Industrial Revolution, 1760–1850." *The West Encounters & Transformations.* Pearson. Retrieved from: http://wps.ablongman.com/long_levack_wc_1/43/11053/2829693.cw/

Lewis, H. T. 1982. A time for burning. University of Alberta Boreal Institute for Northern Studies Occasional Publication 17.

Lewis, Meriwether, William Clark, & Thomas Jefferson. 1893. *History of the Expedition Under the Command of Lewis and Clark: To the Sources of the Missouri River, Thence Across the Rocky Mountains and Down the Columbia River to the Pacific Ocean, Performed During the Years 1804-5-6, Volume 3.* Henry Stevens & Son.

"The Life of Henry Ford." 2013. The Henry Ford. Retrieved from: http://www.hfmgv.org/exhibits/hf/

Livezey, K. B. 2010(a). "Killing Barred Owls to Help Spotted Owls I: a Global Perspective." *Northwestern Naturalist,* 91(2), 107–133.

———. 2010(b). "Killing Barred Owls to Help Spotted Owls II: Implications for Many Other Range-Expanding Species." *Northwestern Naturalist,* 91(3), 251–270.

Lott, M. C. 2011, September 27. "Solyndra – Illuminating Energy Funding Flaws?" *Scientific American.* Retrieved on August 9, 2012 from: http://blogs.scientificamerican.com/plugged-in/2011/09/27/solyndra-illuminating-energy-funding-flaws/

Lovelock, James E. 1988. *The Ages of Gaia: A Biography of Our Living Earth.* New York: Norton.

———. 1991(a). *Gaia: The Practical Science of Planetary Medicine.* Oxford University Press, Incorporated.

———. 1991 (b). *Healing Gaia: Practical Medicine for the Planet.* Harmony Books.

———. 2007. *The Revenge of Gaia: Why the Earth is Fighting Back and How We Can Still Save Humanity.* Penguin Books Limited.

Lucas v. South Carolina Coastal Council. 1992. 505 U.S. 1003. Retrieved from: http://www.lexisnexis.com.dist.lib.usu.edu/hottopics/lnacademic/?

Lueck, D., and Michael, J. A. 2003. "Preemptive Habitat Destruction under the Endangered Species Act." *Journal of Law and Economics*, 46(1), 27–60. Retrieved from: http://graphics8.nytimes.com/images/blogs/freakonomics/pdf/FreakPDF3.pdf

Luther, Linda. 2005, November 16. *The National Environmental Policy Act: Background and Implementation*. CRS Report for Congress. Retrieved from: http://www.fta.dot.gov/documents/Unit1_01CRSReport.pdf

Lyons, Michael. October 1999. "Political Self-Interest and U.S. Environmental Policy." *Natural Resources Journal*, 39: 289.

MacCleery, D. 1994, September 7. "Understanding the Role That Humans Have Played in Shaping America's Forest and Grassland Landscapes." Forest Service Department of Agriculture Newsletter.

Macdonald, Matthew. 2007. "Rapanos v. United States and Carabell v. United States Army Corps of Engineers." *Harvard Environmental Law Review*, 31: 321–332. Retrieved from: http://www.law.harvard.edu/students/orgs/elr/vol31_1/macdonald.pdf

Maczulak, Anne E. 2009. *Cleaning Up the Environment: Hazardous Waste Technology*. Infobase Publishing. Retrieved from: http://books.google.com/books?id=6l3jPZXTczwC&dq=he+1970s+must+absolutely+be+the+years+when+America+pays+its+debt+to+the+past+by+reclaiming+the+purity+of+its+air,+its+waters,+and+our+living+environment&source=gbs_navlinks_s

Mandelker, Daniel R. 2010. "The National Environmental Policy Act: A Review of Its Experience and Problems." *Washington University Journal of Law and Policy*, 32: 293–312.

Mann, Charles C. 2006. *1491: New Revelations of the America's Before Columbus*. New York, New York: Vintage Books.

Manson, Pamela, DJ Summers, and Cathy Mckitrick. 2013, May 27. "Not much life left for Utah graveyards." *The Salt Lake Tribune*. Retrieved from: http://www.sltrib.com/sltrib/news/56321559-78/cemetery-cemeteries-burial-plots.html.csp

Marris, Emma. 2011. *Rambunctious Garden: Saving Nature in a Post-Wild World*. New York: Bloomsbury USA.

Martinez, D. 1993, February 26. *Back to the Future: Ecological Restoration, the Historical Forest, and Traditional Indian Stewardship*. Paper Presented at a Watershed Perspective and Native Plants Conference, Olympia Washington.

Massachusetts, et al., v. Environnemental Protection Agency (EPA) et al. 2007. 549 U.S. 497. Retrieved October 21, 2013 from: http://supreme.justia.com/cases/federal/us/549/05-1120/opinion.html

McCann, Kevin S. 2012. *Food Webs*. Princeton: Princeton University Press.

McCarthy, Michael. 2012, June 12. "Earthrise: The image that changed our view of the planet." *The Independent*. Retrieved from: http://www.independent.co.uk/environment/nature/earthrise-the-image-that-changed-our-view-of-the-planet-7837041.html

McChesney, Robert W. 1997. *Corporate Media and the Threat to Democracy*. Seven Stories Press.

McClure, Robert, and Bonnie Stewart. 2012, July 18. *Clean Water Act's Anti-Pollution Goals Prove Elusive.* Oregon Public Broadcasting. Retrieved from: http://earthfix.opb.org/water/article/anti-pollution-goals-elude-clean-water-act-enforce/

McInnis, Doug. Summer 2006. "What System?" Dartmouth Medicine. Retrieved from: http://dartmed.dartmouth.edu/summer06/html/what_system_03.php

McKibben, Bill. 1989. *The End of Nature.* New York: Random House Inc.

McKinney, Larry. 1993. "Reauthorizing the Endangered Species Act—Incentives for Rural Landowners." In *Building Economic Incentives into the Endangered Species Act: A Special Report from Defenders of Wildlife*, edited by Wendy E. Hudson. Washington, D.C.: Defenders of Wildlife.

McLendon, Russell. 2012(a), March 13. Endangered Species Act Stirs New Debates. *Mother Nature Network.* Retrieved from: http://www.mnn.com/earth-matters/politics/blogs/endangered-species-act-stirs-new-debates

———. 2012(b), September 11. *Clean Water Act Is 40 Years Old: Landmark Water Law Hits a Milestone During Critical Time.* The Huffington Post. Retrieved from: http://www.huffingtonpost.com/2012/09/11/clean-water-act-2012_n_1874980.html

Meadows, Donella H., Club of Rome, Potomac Associates. 1972. *The Limits to Growth.* Universe Books.

Meiners, Roger, and Bruce Yandle. n.d. *The Common Law: How it Protects the Environment.* Property and Environment Research Center (PERC). Retrieved from: http://perc.org/articles/common-law-0

———. June 1998. "How the Common Law Protects the Environment: Curbing Pollution–Case-By-Case." *PERC Reports*, 16: 7–9. Retrieved from: http://perc.org/sites/default/files/june98.pdf

Michael, J. A. 2000, August. The Endangered Species Act and Private Landowner Incentives. *United States Department of Agriculture: Animal and Plant Health Inspection Service*, p. 29. Retrieved from: http://www.aphis.usda.gov/wildlife_damage/nwrc/symposia/economics_symposium/michael.HR.pdf

Miller, Char. 2012. *Public Lands, Public Debates: A Century of Controversy.* Oregon University Press. Retrieved from: http://osupress.oregonstate.edu/sites/default/files/Miller.PublicLandsPublicDebates.Excerpt.pdf

Miller, Henry I., and Gregory Conko. 2012, September 5. "Rachel Carson's Deadly Fantasies." *Forbes.* Retrieved from: http://www.forbes.com/sites/henrymiller/2012/09/05/rachel-carsons-deadly-fantasies/

Mills, L. Scott, Michael E. Soulé, and Daniel F. Doak. 1993. "The Keystone-Species Concept in Ecology and Conservation." *Bioscience*, 43: 219–224.

Mills, Stephanie. 1995. *In Service of the Wild.* Boston: Beacon Press.

Morath, S. J. 2008. "Endangered Species Act: A New Avenue for Climate Change Litigation." *Public Land & Resources Law Review*, 29: 23–40. Retrieved from: http://scholarship.law.umt.edu/cgi/viewcontent.cgi?article=1253&context=plrlr.

Morrison, J. 2004, August 14. "Shoot, Shovel & Shut-Up." *NewsWithViews*. Retrieved from: http://www.newswithviews.com/Morrison/joyce7.htm

Morriss, Andrew P. 2000. "The Politics of the Clean Air Act." *Political Environmentalism*. Edited by Terry L. Anderson. Stanford: Hoover Institution Press.

Mountain Democrat. 2012. "Gas fracking." Retrieved August 2, 2012 from: http://www.mtdemocrat.com/opinion/gas-fracking/

Mueller, John E. 1999. *Capitalism, Democracy, and Ralph's Pretty Good Grocery*. Princeton, NJ: Princeton University Press.

Murchison, K. M. 2007, March 1. The Snail Darter Case. *University Press of Kansas*. Retrieved from: http://www.kansaspress.ku.edu/mursna.html

Murray, Ann. 2009. *Smog Deaths In 1948 Led To Clean Air Laws*. NPR. Retrieved from: http://www.npr.org/templates/story/story.php?storyId=103359330

Nardinelli, Clark. 2008. "Industrial Revolution and the Standard of Living." *Library of Economics and Liberty*. Retrieved from: http://www.econlib.org/library/Enc/IndustrialRevolutionandtheStandardofLiving.html

Nash, Roderick F. 2001. *Wilderness and the American Mind*. 4th ed. 1967. Reprint, New Haven: Yale University Press.

National Archives. U.S. National Archives and Records Administration. n.d. *Teaching With Documents: The Homestead Act of 1862*. Retrieved from: http://www.archives.gov/education/lessons/homestead-act/

National Atlas of the United States. 2003. *Federal Lands and Indian Reservations: Utah*. Retrieved from: http://nationalatlas.gov/printable/fedlands.html#ut

National Environmental Policy Act. 1969. U.S. Code. Title 42. Sections 4321–4347.

National Oceanic and Atmospheric Administration. n.d. *Full Text of the Endangered Species Act (ESA)*. Retrieved from: www.nmfs.noaa.gov/pr/laws/esa/text.htm#section3

National Park Service. 2007. *Lamar Valley and Wolves*. Retrieved from: http://www.nps.gov/features/yell/insideyellowstone/lamarvalleytranscript.htm

———. 2013, September 5. Theodore Roosevelt and Conservation. U.S. Department of Interior. Retrieved from: http://www.nps.gov/thro/historyculture/theodore-roosevelt-and-conservation.htm

National Wildlife Federation. n.d. Keeping the Endangered Species Act Strong. *National Wildlife Federation*. Retrieved from: https://www.nwf.org/What-We-Do/Protect-Wildlife/Endangered-Species/Endangered-Species-Act.aspx

Natural Resources Conservation Service. Department of Agriculture. March 2011. *Assessment of the Effects of Conservation Practices on Cultivated Cropland in the Chesapeake Bay Region*. Retrieved from: http://www.nrcs.usda.gov/Internet/FSE_DOCUMENTS/stelprdb1042077.pdf

Natural Resources Defense Council. n.d. Court Upholds Endangered Species Act Protection for Polar Bears. *Natural Resources Defense Council*. Retrieved from http://www.nrdc.org/media/2011/110630.asp

Nature Conservancy. 2011, February 14. "Red-cockaded Woodpecker species profile." *Nature Conservancy*. Retrieved from: http://www.nature.org/ourinitiatives/regions/northamerica/unitedstates/mississippi/explore/red-cockaded-woodpecker-species-profile.xml

Neel, M. C., Leidner, A. K., Haines, A., Goble, D. D., & Scott, J. 2012. By the Numbers: How is Recovery Defined by the US Endangered Species Act? *Bioscience*, 62(7), 646–657. doi:10.1525/bio.2012.62.7.7

Neimark, Peninah, and Peter Rhoades Mott. 2011. *The Environmental Debate: A Documentary History, with Timeline, Glossary, and Appendices.* 2nd ed. Amenia, NY: Grey House.

Nelson, Robert H. 1995. *Public Lands and Private Rights: The Failure of Scientific Management.* Lanham: Rowman & Littlefield.

———. 2010. The *New Holy Wars: Economic Religion vs. Environmental Religion in Contemporary America.* Pennsylvania State University Press for Independent Institute.

Nijhuis, M. 2012. "Which Species Will Live?" *Scientific American*, 307(2), 74–79.

Norton v. Southern Utah Wilderness Alliance. 2004. 542 U.S. 55. Retrieved from: http://www.leagle.com/decision/2004597542US55_1596

Norton, Helen H. 1979. "The association between anthropogenic prairies and important plants in western Washington." *Northwest Anthropological Research Notes*, 13: 175–200.

No Wolves. n.d. Anti Wolf Tshirts. *No Wolves*. Retrieved from: http://www.nowolves.com/nowolves/T-Shirts/default.html

Oates, Wallace E. November 2001. *A Reconsideration of Environmental Federalism*. Resources for the Future. Retrieved from: http://www.rff.org/documents/rff-dp-01-54.pdf

———. Spring 2002. "The Arsenic Rule: A Case in Decentralized Standard Setting?" *Resources for the Future*, 147: 16–18. Retrieved from: http://www.rff.org/rff/documents/rff-resources-147-arsenicrule.pdf

Oblack, Rachelle. n.d. "What is Smog?" About.com. Retrieved from: http://weather.about.com/od/ozoneinformation/qt/smogcity.htm

O'Connor, M. C. 2013, February 28. Can the Endangered Species Act Protect Against Climate Change? *Outside*. Retrieved from: http://www.outsideonline.com/adventure-travel/adventure-ethics/Rising-Seas-and-the-Endangered-SpeciesAct.html?page=1

O'Donoghue, Amy Joi, and Emiley Morgan. 2011, March 4. "Jury finds activist Tim DeChristopher guilty of both charges." *Deseret News*. Retrieved from: http://www.deseretnews.com/article/705367933/Jury-finds-activist-Tim-DeChristopher-guilty-of-both-charges.html?pg=all

Office of Wetlands, Oceans, & Watersheds. November 2011. *A National Evaluation of the Clean Water Act Section 319 Program*. Retrieved from: http://www.epa.gov/owow/NPS/pdf/319evaluation.pdf (From Chapter 9: Clean Water Act)

Ogden, Peter S. 1950. "Peter Skene Ogden's Snake Country Journals." 1824–25 and 1825–26. E. E. Rich & A. M. Johnson (Eds.). Hudson's Bay Record Society Publication 8.

O'Keefe, M. T. n.d. Red-Cockaded Woodpeckers. *Florida Wildlife Viewing*. Retrieved from: http://www.floridawildlifeviewing.com/florida_animals_wildlife/Cockaded Woodpecker.html

Oregon Fish and Wildlife Office, U.S. Fish and Wildlife Service. 2012, November 5. *Northern Spotted Owl*. Retrieved from: www.fws.gov/oregonfwo/species/data/northern spottedowl/

O'Toole, R. 2003, July 1. "Census Bureau: 94.6% of U.S. Is Rural Open Space." *Heartlander Magazine*. Retrieved from: http://news.heartland.org/newspaper-article/2003/07/01/census-bureau-946-percent-us-rural-open-space

"Our Voice: Bladderpod controversy reinforces need to reform Endangered Species Act." *Tri-City Herald*. 2013, August 11. Retrieved from: http://www.tri-cityherald.com/2013/08/11/2514511/our-voice-bladderpod-controversy.html

"Paul Müller-Biographical." n.d. The Nobel Foundation. Retrieved from: http://www.nobelprize.org/nobel_prizes/medicine/laureates/1948/muller-bio.html

Pederson, William F., Jr. May 1981. "Why the Clean Air Act Works Badly." *University of Pennsylvania Law Review*, 129: 1059, 1109. Retrieved October 27, 2013 from: http://www.jstor.org/stable/3311951

Peterson, Jodi. n.d. *The end of 'analysis paralysis'?* High Country News. Retrieved from: http://www.hcn.org/issues/340/16838/print_view

Pielou, Evelyn C. 1991. *After the Ice Age: The Return of Life to Glaciated North America*. Chicago: University of Chicago Press.

Plater, Zymunt J. B. 1986, July 1. "In the Wake of the Snail Darter: An Environmental Law Paradigm and Its Consequences." *Journal of Law Reform*, Vol. 19:4 Summer 1986, p. 813. Retrieved from: http://lawdigitalcommons.bc.edu/cgi/viewcontent.cgi?article=1329&context=lsfp

———. 2013. *The Snail Darter and the Dam: How Pork-Barrel Politics Endangered a Little Fish and Killed a River*. New Haven: Yale University Press.

PLOS Medicine. 2(8): e124. Retrieved from: http://www.plosmedicine.org/article/info:doi/10.1371/journal.pmed.0020124

Plumer, Brad. 2011, September 30. "Energy secretary Chu felt pressure to speed up loans to firms like Solyndra." *Washington Post*. Retrieved from http://www.washingtonpost.com/business/economy/energy-secretary-chu-felt-pressure-to-speed-up-loans-to-firms-like-solyndra/2011/09/30/gIQAXhXHBL_story.html

———. 2013, June 25. "Obama may have left himself wiggle room to approve Keystone XL." *Washington Post*. Retrieved from: http://www.washingtonpost.com/blogs/wonkblog/wp/2013/06/25/did-obama-leave-himself-some-wiggle-room-to-approve-keystone-xl/

Pociask, Steve, & Joseph Fuhr. 2011, March 10. *Progress Denied: A Study on the Potential Economic Impact of Permitting Challenges Facing Proposed Energy Projects*. Prepared for U.S. Chamber of Commerce.

Powers, Ann. 2004. *Federal Water Pollution Control Act (1948) (Major Acts of Congress)*. http://www.enotes.com/topics/federal-water-pollution-control-act-1948/reference#reference-federal-water-pollution-control-act-1948

Pritchett, L. 2006. "Sight the Gun High." *Natural Resources Journal*, 46(1), 1–8.

Profeta, T. H. 1996. *Managing without a balance: Environmental regulation in light of ecological advances*. Duke Environmental Law and Policy Forum, 7, 71–103.

Profita, Cassandra. 2012, July 30. *Oregon Farmers Go Beyond The Clean Water Act*. Oregon Public Broadcasting. Retrieved from: http://www.opb.org/news/blog/ecotrope/oregon-farmers-go-beyond-the-clean-water-act/

Public Land Law Review Commission, Bureau of Land Management. 1970, June. *One Third of the Nation's Land: A Report to the President and to the Congress by the Public Land Law Review Commission*. Washington DC: Government Printing Office. Retrieved from: http://ia601500.us.archive.org/34/items/onethirdofnation3431unit/onethirdofnation3431unit.pdf

Ramenofsky, Ann F. 1987. *Vectors of Death: The Archaeology of European Contact*. Albuquerque, NM: University of New Mexico Press.

Rapanos v. United States. 2006. 547 U.S. 715. Retrieved from: http://www.law.cornell.edu/supct/html/04-1034.ZO.html

Rasker, Ray, Norma Tirrell, Deanne Klopfer. 1992. *The Wealth of Nature: New Economic Realities in the Yellowstone Region*. Washington, DC: Wilderness Society.

Ray, Charles Michael. 2010, September 9. "Forest Near Mount Rushmore Suffers Beetle Attack." NPR. Retrieved from: http://www.npr.org/templates/story/story.php?storyId=130053576

Raynolds, F. W. 1868. *Report on the Exploration of the Yellowstone River in 1859–60*. Senate Executive Document 77, 40th Congress, Second Session.

Reason Foundation. 2007, June 27. "Bald Eagle Off Endangered List In Spite of Feds, Not Because of Them." *Reason Foundation*. Retrieved from: http://reason.org/news/show/1002874.html

Reiland, R. R. 2004, April 7. "Shoot, Shovel & Shut Up." *The American Spectator*. Retrieved from: http://spectator.org/archives/2004/04/07/shoot-shovel-shut-up/print

Rein, W., Andreas, E., and Kennedy, K. 2012, June 28. "NEPA and renewable energy practices: streamlining sustainability." *Association of Corporate Counsel*. Retrieved from:" http://www.lexology.com/library/detail.aspx?g=6c2f0e4a-48a9-4419-95f3-eb6890fb0b10

Restore Hetch Hetchy. n.d. "Hetch Hetchy Timeline." Retrieved from: http://www.hetchhetchy.org/originalvalley/timeline

Restuccia, Andrew. 2012, June 28. "Another Solyndra? DOE: Loan recipient closing." *Politico*. Retrieved from: http://www.politico.com/news/stories/0612/77966.html

Revkin, A. C. 2009, May 8. "U.S. Curbs Use of Species Act in Protecting Polar Bear." *New York Times*. Retrieved from: http://www.nytimes.com/2009/05/09/scienceearth/09bear.html?_r=0

Reynolds, Andy. n.d. "A Brief History of Environmentalism." Retrieved from: http://webcache.googleusrcontent.com/search?q=cache:0ZFYRS7Jtt4J:www.public.iastate.edu/~sws/enviro%2520and%2520society%2520Spring%25202006/Historyof Environmentalism.doc+&cd=1&hl=en&ct=clnk&gl=us

Riccardi, Nathan D. 2012. "Necessarily Hypocritical: The Legal Viability of EPA's Regulation of Stationary Source Greenhouse Gas Emissions Under the Clean Air Act." *Boston College Environmental Affairs Law Review*, 39(1), 213–241.

Richards, K. R. n.d. Biodiversity and Species Protection. *The School of Public and Environmental Affairs*. Retrieved from: classwebs.spea.indiana.edu/kenricha/biodiversity/esa.htm

Richards, Mary. 2012, November 12. "Group to ask President Obama to turn Greater Canyonlands into national monument." *Deseret News*. Retrieved from: http://www.deseretnews.com/article/865566603/Group-to-ask-President-Obama-to-turn-Greater-Canyonlands-into-national-monument.html?pg=all

Ridenour, David. 2005, August. "TESRA" Endangered Species Act Reform Proposal Would Do More Harm Than Good. *National Center for Public Policy Research*. Retrieved from: http://www.nationalcenter.org/NPA531TESRA.html

Ring, Ray. n.d. "Supreme Court reins in citizens' right to sue." *High Country News*. Retrieved from: http://www.hcn.org/issues/278/14856/print_view

Robinson, S. 2010. *Comments on the EPA Draft Chesapeake Bay TMDL*. National Association of Conservation Districts. Retrieved from: http://www.nacdnet.org/doc_details/280-chespeake-bay-tmdl

Rogers, S. n.d. "11 Bizarre Sources of Clean Energy, from Dead Turkeys to Urine." Retrieved July 11, 2012 from: http://web.archive.org/web/20100419083510/http:/earthfirst.com/11-bizarre-sources-of-clean-energy-from-dead-turkeys-to-urine/

Romboy, Dennis. 2012, March 4. "Utah poised to lead another Sagebrush Rebellion over federal land." *Deseret News*. Retrieved from: http://www.deseretnews.com/article/865551474/Utah-poised-to-lead-another-Sagebrush-Rebellion.html?pg=all

Rosenberg, A. 2013, August 19. "Wolves, the Endangered Species Act, and Why Scientific Integrity Matters." *The Equation*. Retrieved from: http://blog.ucsusa.org/wolves-the-endangered-species-act-and-why-scientific-integrity-matters-212

Rosenblatt, Joel. 2012, July 27. "Solyndra files plan to reorganize in chapter 11 bankruptcy." *Bloomberg*. Retrieved from: http://www.bloomberg.com/news/2012-07-28/solyndra-files-plan-to-reorganize-in-chapter-11-bankruptcy.html

Rotman, Michael. n.d. "Cuyahoga River Fire." *Cleveland Historical*. Retrieved from: http://clevelandhistorical.org/items/show/63#.Um19FhAUYSk

Rubin, Charles T. 1994. *The Green Crusade*. New York: The Free Press.

Rural Liberty Alliance. 2011, December 6. "The ESA: How Litigation Is Costing Jobs and Impeding Recovery." Retrieved from: http://ruralliberty.org/the-endangered-species-act-how-litigation-is-costing-jobs-and-impeding-true-recovery-efforts

Russell, Carl P. 1941. "Coordination of Conservation Programs." *Regional Review*, 6: (1-2): 13–19. Retrieved from: http://www.cr.nps.gov/history/online_books/regional_review/vol6-1-2e.htm

"The Sagebrush Rebellion." 1980, December 1. *U.S. News and World Report*. Retrieved from: http://www2.vcdh.virginia.edu/PVCC/mbase/docs/sagebrush.html

Salter, Trevor. Spring 2011. "NEPA and Renewable Energy: Realizing the most Environmental Benefit in the Quickest Time." *Environs*, 34: 173–187.

Scheer, Roddy, and Doug Moss. 2012, February 26. "Why did solar panel maker Solyndra fail?" *E – The Environmental Magazine*. Retrieved from: http://www.emagazine.com/earth-talk/why-did-solar-panel-maker-solyndra-fail/

Schindler, D. W. 1976, May 7. "The Impact Statement Boondoggle." *Science*. 192: 509.

Schullery, Paul. 2004. *Searching for Yellowstone: Ecology and Wonder in the Last Wilderness*. Helene, MT: Montana Historical Society Press.

Scientific American. 2012, August 9. "Is the Endangered Species Act a Success or Failure?" *Scientific American*. Retrieved from: http://www.scientificamerican.com/article.cfm?id=endangered-species-act-success-failure

Scott, Douglas W. June 2001, rev. November 2003. "A Wilderness-Forever Future: A Short History of the National Wilderness Preservation System." http://wilderness.nps.gov/celebrate/Section_Two/NWPS%20History.pdf

Seasholes, B. 2007(a), June 1. The Bald Eagle, DDT, and the Endangered Species Act: Examining the Bald Eagle's Recovery in the Contiguous 48 States. *Reason Foundation*. Retrieved from: http://reason.org/news/show/the-bald-eagle-ddt-and-the-end

———. 2007(b), June 27. Reason Foundation—Bald Eagle Off Endangered List in Spite of Feds, Not Because of Them. *Reason Foundation*. Retrieved from: http://reason.org/news/show/1002874.html

Semple, Robert B. 2012, October 16. "Happy Birthday, Clean Water Act." *New York Times Opinion Pages*. Retrieved from: http://takingnote.blogs.nytimes.com/2012/10/16/happy-birthday-clean-water-act/

Sexton, Ken. 1995. "Science and Policy in Regulatory Decision Making: Getting the Facts Right about Hazardous Air Pollutants." *Environmental Health Perspectives*, 103: 213, 222. Retrieved on October 28, 2013 from: http://www.jstor.org/stable/3432539

Shabecoff, Philip. 1993. *A Fierce Green Fire: The American Environmental Movement*. New York: Hill and Wang.

Sherman, Brad. 2006. "A Time to Act Anew: A Historical Perspective on the Energy Policy Act of 2005 and the Changing Electrical Energy Market." *William & Mary Environmental Law and Policy Review*, 31(1). Retrieved from: http://scholarship.law.wm.edu/cgi/viewcontent.cgi?article=1095&context=wmelpr

Shetler, Jan B. 2007. *Imagining Serengeti: A History of Landscape Memory in Tanzania from Earliest Times to the Present*. Athens: Ohio University Press.

"Ship Subsidies." 1908, April 11. The Outlook. Retrieved on August 6, 2012 from: http://www.unz.org/Pub/Outlook-1908apr11-00815

Shumacher, Ernst F. 1973. *Small is Beautiful: A Study of Economics as if People Mattered*. Blond and Briggs.

Simmons, Randy T. 2012. *Beyond Politics: The Roots of Government Failure*. Independent Publishing Group.

Simmons, Randy T and Ryan M. Yonk. 2013. "Economics Impacts of Southern Utah Wilderness Alliance Litigation on Local Communities." Retrieved from: http://www.usu.edu/ipe/wp-content/uploads/ipePublications/Economic-Impacts-of-Southern-Utah-Wilderness-Alliance-on-Lcoal-Comminities.pdf

Slootmaker, Stelle. 2011. "John Muir: Nature's Evangelist?" Retrieved from http://griid.org/2011/01/21/john-muir-natures-evangelist/

Smillie, Susan M., & Lucinda Low Swartz. 2002. *Achieving the 150-Page Environmental Impact Statement.* Retrieved from: http://www.lucindalowswartz.com/images/150_Pg_EIS.pdf

Smith, Anne E. December 2011. "An Evaluation of the $PM_{2.5}$ Health Benefits Estimates in Regulatory Impact Analyses for Recent Air Regulations." NERA Economic Consulting. Retrieved from: http://www.nera.com/nera-files/PUB_RIA_Critique_Final_Report_1211.pdf

Smith, Marvin T. 1987. *Archaeology of Aboriginal Culture Change in the Interior Southeast.* Gainesville, FL: Florida State Museum.

Snyder, Jim. 2012, August 2. Solyndra report ending U.S. House probe fuels campaign rhetoric. *Bloomberg Businessweek.* Retrieved from: http://www.businessweek.com/news/2012-08-02/solyndra-report-ending-u-dot-s-dot-house-probe-fuels-campaign-rhetoric

Solid Waste Agency v. Army Corps of Engineers. 2001. 531 U.S. 159. Retrieved from: http://www.oyez.org/cases/2000-2009/2000/2000_99_1178

Southern Utah Wilderness Alliance v. Norton. 2002(a). 301 F.3d 1217. Retrieved from: http://scholar.google.com/scholar_case?case=17871977932500995166&q=suwa&hl=en&as_sdt=2,45

———. 2002(b). The Oyez Project at IIT Chicago-Kent College of Law. Retrieved from: http://www.oyez.org/cases/2000-2009/2003/2003_03_101

Southern Utah Wilderness Alliance. 2013(a). "About SUWA." Retrieved from: http://www.suwa.org/about-suwa/

———. 2013(b). "Protecting Greater Canyonlands." Retrieved from: http://www.suwa.org/issues/greatercanyonlands/

———. 2014. "County Official Defies Federal Agency Authority in Recapture Canyon." Retrieved from: http://www.suwa.org/2014/05/13/county-official-defies-federal-agency-authority-in-recapture-canyon/

Species Profile. U.S. Fish and Wildlife Service Home. Retrieved from: http://www.fws.gov/arcata/es/birds/nso/ns_owl.html

Spitzer, Nick. 2012, February 15. "The Story of Woody Guthrie's 'This Land Is Your Land.'" *NPR Music.* Retrieved from: http://www.npr.org/2000/07/03/1076186/this-land-is-your-land

Stannard, David E. 1992. *American Holocaust: Columbus and the Conquest of the New World.* New York: Oxford University Press.

Stearn, Esther W., Allen E. Stearn. 1945. *The Effect of Smallpox on the Destiny of the Amerindian.* Boston, MA: Bruce Humphries, Inc.

Stevenson, Merrill. 2010, March 1. The Odd Limits of the Clean Water Act. *The Economist*. Retrieved from: http://www.economist.com/blogs/democracyinamerica/2010/03/congress_and_clean_water_act

Stolovitsky, Neil. 2011, July 20. "The Model T Ford project: a model for product and process innovation." Retrieved Afrom: http://pmbox.geniusinside.com/manufacturing/the-model-t-ford-project-a-%E2%80%9Cmodel%E2%80%9D-for-product-and-process-innovation/

Stone, Robert. 2009. *Earth Days. USA: American Experience*. Film.

Story of Wild Horse Annie. Retrieved from: http://www.ispmb.org/AnniesStory

Strain, D. 2011, July 28. "House Strikes Proposed Ban on Endangered Species Listings." *ScienceInsider*. Retrieved from: http://news.sciencemag.org/scienceinsider/2011/07/house-strikes-proposed-ban-on.html

Stroll, Ted. 2010, August, 26. "Aw, Wilderness!" *New York Times*. Retrieved from: http://www.nytimes.com/2010/08/27/opinion/27stroll.html?_r=0

"Stuart Udall Biography." 2003, November 14. NOW with Bill Moyers. Science and Health: Preserving the Parks. PBS. Retrieved from: http://www.pbs.org/now/science/udall.html

Suckling, K. 2011, December 2. Testimony on "The Endangered Species Act: How Litigation Is Costing Jobs and Impeding True Recovery Efforts." *Center for Biological Diversity*. Retrieved from: http://www.biologicaldiversity.org/campaigns/esa_works/pdfs/SucklingKieran_Written_Testimony_12-6-11.pdf

Suckling, K., Greenwald, N., & Curry, T. 2012, May 1. On Time, On Target: How the Endangered Species Act is Saving America's Wildlife. *Center for Biological Diversity*. Retrieved from: www.esasuccess.org/pdfs/110_REPORT.pdf

SummitPost. n.d. Public and Private Land Percentages by US States. *SummitPost*. Retrieved from: http://www.summitpost.org/public-and-private-land-percentages-by-us-states/186111

Tabb, William Murray. 1997. "The Role of Controversy in NEPA: Reconciling Public Veto with Public Participation in Environmental Decisionmaking." *William & Mary Environmental Law and Policy Review*, 21: 175–231.

Take Part. 2013. "An Inconvenient Truth: The Official Website of the Award-Winning Film." Retrieved from: http://www.takepart.com/an-inconvenient-truth/film

Taylor Grazing Act. 1934. *U.S. Code*. Title 43. Section 315.

Taylor, Joseph E., III, and Matthew Klingle. 2006, March 9. "Environmentalism's Elitist Tinge has Roots in the Movement's History." *Grist*. Retrieved from: http://grist.org/article/klingle/

Teensma, Peter, Dominic Adrian, John T. Rienstra, and Mark A. Yeiter. 1991. "Preliminary Reconstruction and Analysis of Change in Forest Stand Age Classes of the Oregon Coast Range From 1850 to 1940." *USDAI Bureau of Land Management Technical Note*, T/N OR-9.

Tennessee Valley Authority (TVA) v. Hill et al. 1978. 437 U.S. 153. Retrieved from: www.lexisnexis.com/hottopics/lnacademic

Tennessee Valley Authority (TVA). n.d. Frequently Asked Questions About TVA. *Tennessee Valley Authority.* Retrieved from: http://www.tva.com/abouttva/keyfacts.htm

Theodore Roosevelt Association. 2013. "Brief Biography." Retrieved from: http://www.theodoreroosevelt.org/site/c.elKSIdOWIiJ8H/b.8684643/k.9DB0/Brief_Biography.htm

Theodore Roosevelt Conservation Partnership v. Salazar. 2011. 661 F.3d 66. Retrieved from: http://www.law.indiana.edu/publicland/files/trcp_v_salazar_edited.pdf

Thoreau, Henry D. 1910. *Walden.* New York: Thomas Y. Crowell & Company.

Tierney, John. 2013, September 16. "The Rational Choices of Crack Addicts." *New York Times.* Retrieved from: http://www.nytimes.com/2013/09/17/science/the-rational-choices-of-crack-addicts.html?pagewanted=all&_r=0

"Tim DeChristopher." 2013, March 1. "The New Political Prisoners: Leakers, Hackers and Activists." *Rolling Stone Politics.* Retrieved from: http://www.rollingstone.com/politics/lists/the-new-political-prisoners-leakers-hackers-and-activists-20130301/tim-dechristopher-19691231

Trafton, Anne. 2009. "New virus-built battery could power cars, electronic devices." MIT News. Retrieved October 26, 2013 from: http://mitei.mit.edu/news/new-virus-built-battery-could-power-cars-electronic-devices

Turner, Frederick Jackson. "The Significance of the Frontier in American History." American Historical Association. Chicago Worlds Fair. Chicago, 12 July 1893. E 179.5 .T958 1966

Turner, James M. 2009. "The Specter of Environmentalism: Wilderness, Environmental Politics, and the Evolution of the New Right." *Journal of American History,* 96: 123–148.

―――. 2012. *The Promise of Wilderness: American Environmental Politics Since 1964.* Seattle: University of Washington.

Turner, N.J. 1991. "Burning mountain sides for better crops: Aboriginal landscape burning in British Columbia." *Archaeology in Montana,* 32: 57–73.

Udall, Stewart L. 1963. *The Quiet Crisis.* New York: Holt, Rinehart and Winston.

University of California, Santa Barbara, Department of Geography. 2010, May 27. *Keith Clarke Quoted by the BBC Re Offshore Drilling.* Retrieved from: http://www.geog.ucsb.edu/events/department-news/713/keith-clarke-quoted-by-the-bbc-re-offshore-drilling/

University of Nevada. 2008, June 22. *Special Collections: A Guide to the Records of Sagebrush Rebellion: Collection No. 85-04.* Retrieved from: http://knowledgecenter.unr.edu/specoll/mss/85-04.html

U.S.-Canada Power System Outage Task Force. 2006, September. Final Report on the Implementation of the Task Force Recommendations. Retrieved from: http://energy.gov/sites/prod/files/oeprod/DocumentsandMedia/BlackoutFinalImplementationReport%282%29.pdf

U.S. Census Bureau. 2010. Fruit Heights, Population. Retrieved from: https://www.google.com/search?q=fruit+heights+population&oq=fruit+heights+population&aqs=chrome..69i57j0l2.32349j0&sourceid=chrome&ie=UTF-8

U.S. Congress. 1977. *Congressional Record*. 95th Congress, 1st Session. Retrieved on October 27, 2013 from: http://abacus.bates.edu/muskiearchives/ajcr/1977/CAA%20Metzenbaum.shtml

U.S. Department of Energy. 2013, September 6. *National Environmental Policy Act: Lessons Learned*. Retrieved from: http://energy.gov/sites/prod/files/2013/09/f2/LLQR-2013-Q3.pdf

U.S. Fish and Wildlife Service. 2009, June. *Our Endangered Species Program and How It Works with Landowners*. Retrieved from: http://www.fws.gov/endangered/esa-library/pdf/landowners.pdf

⸻. 2013(a), January. *ESA Basics: 40 Years of Conserving Endangered Species*. Retrieved from: http://www.fws.gov/endangered/esa-library/pdf/ESA_basics.pdf

⸻. 2013(b), July 15. *Endangered Species Program Timeline*. Retrieved from: http://www.fws.gov/endangered/laws-policies/timeline.html

⸻. 2013(c), July. *1982 ESA Amendment*. Retrieved from: http://www.fws.gov/endangered/laws-policies/esa-1982.html

U.S. Forest Service. United States Department of Agriculture. 2013. *Landmark Restoration Effort to Heal Northern Arizona's Forests and Communities*. Retrieved from: https://fs.usda.gov/Internet/FSE_DOCUMENTS/stelprdb5158926.pdf

⸻. n.d. *Wilderness Management 101*. Retrieved from: http://www.fs.usda.gov/Internet/FSE_DOCUMENTS/fsbdev3_053246.pdf

U.S. Government Printing Office. n.d(a). 16 U.S.C. § 1531. Retrieved from: www.gpo.gov/fdsys/pkg/USCODE-2011-title16/pdf/USCODE-2011-title16-chap35-sec1531.pdf

⸻. n.d(b). 16 U.S.C. § 1532. Retrieved from: http://www.gpo.gov/fdsys/pkg/USCODE-2011-title16/pdf/USCODE-2011-title16-chap35-sec1532.pdf

U.S. House of Representatives. Committee on Resources. 2005, November 17. *House Oversight Hearing*. 109th Congress. Retrieved from: http://www.gpo.gov/fdsys/pkg/CHRG-109hhrg24682.htm

U.S. House of Representatives. 2012, August 2. *The Solyndra Failure: Majority Staff Report*. Retrieved from: energycommerce.house.gov/Media/file/PDFs/Solyndra/solyndrareport.pdf

⸻. 2012. *Rapid Act*. 105th Congress, 2nd session Rept. 112–596. Retrieved from: http://thomas.loc.gov/cgi-bin/cpquery/T?&report=hr596p1&dbname=112&

⸻. 2013. *Fruit Heights Land Conveyance Act*. 113th Congress, 1st session, H933.

U.S. Supreme Court Media. 2013. *Massachusetts v. Environmental Protection Agency*. Oyez: IIT Chicago-Kent College of Law. Retrieved October 18, 2013 from: http://www.oyez.org/cases/2000-2009/2006/2006_05_1120/

Vines, J., S. Salek, & K. Desloover. 2012, December. *Reforming NEPA Review of Energy Projects.* King & Spalding. Retrieved from: http://www.kslaw.com/library/newsletters/EnergyNewsletter/2012/December/article1.html

Wald, Matthew L. 2011, October 6. "E-mails suggest White House weighed a 2nd Solyndra loan worth almost half a billion dollars." *New York Times.* Retrieved from: http://www.nytimes.com/2011/10/06/us/politics/2nd-us-loan-to-solyndra-said-to-have-been-considered.html

Wald, Matthew L. & Michael Kanellos. 2011, September 8. "F.B.I. Raids Solar Firm That Got U.S. Loans." *New York Times.* Retrieved from: http://www.nytimes.com/2011/09/09/business/solar-company-is-searched-by-fbi.html?ref=solyndra

Wang, Ucilia. 2008, October 7. Solyndra rolls out tube-shaped thin film. *Greentechsolar.* Retrieved from: http://www.greentechmedia.com/articles/read/solyndra-rolls-out-tube-shaped-thin-film-1542/

Weber, Max. 1947. *The Theory of Social and Economic Organization.* Translated by A.M. Henderson and Talcott Parsons. Edited with an Introduction by Talcott Parsons: New York Free Press.

Weissman, Jordan. 2012, July 9. "Another Solyndra in the making?" *The Atlantic.* Retrieved from: http://www.theatlantic.com/business/archive/2012/07/another-solyndra-in-the-making/259573/#

Welch, L. A. 1994. "Property Rights Conflicts Under the Endangered Species Act: Protection of the Red Cockaded Woodpecker." In B. Yandle (Ed.), *Land Rights: The 1990's Property Rights Rebellion.* Lanham, MD: Rowman and Littlefield Publishers, Inc.

Western Energy Alliance. n.d. *Government Delays Preventing Jobs and Economic Growth.* Retrieved from: http://www.westernenergyalliance.org/printpdf/481

Wetzler, A. 2007, August 24. "Should We Pay Landowners not to Kill Endangered Species?" *Natural Resources Defense Council.* Retrieved from: http://switchboard.nrdc.org/blogs/awetzler/should_we_pay_landowners_not_t.html

"What is 'Wilderness'?" n.d. Wilderness.net. Retrieved from: http://www.wilderness.net/nwps/whatiswilderness

White, Kathleen Hartnett. May 2012. "EPA's Pretense of Science: Regulating Phantom Risks." Texas Public Policy Foundation. Retrieved on October 28, 2013 from: http://www.texaspolicy.com/sites/default/files/documents/epa-pretense-of-science-acee-kathleen-hartnett-white.pdf

Wilcove, David S., Michael J. Bean, Robert Bonnie, and Margaret McMillan, "Rebuilding the Ark, Toward a More Effective Endangered Species Act for Private Land," December 5, 1996, *On-Line Learning at the Cumberland School of Law.* Retrieved from: omnilearn.net/esacourse/pdfs/Rebuilding_the_Ark.pdf

Wild Free-Roaming Horses and Burros Act. 1971. U.S. Code. Title 16. Section 1331 et seq.

The Wilderness Act. 1964. U.S. Code. Title 16. Sections 1131–1136. Retrieved from: http://wilderness.nps.gov/document/WildernessAct.pdf

Wilderness Society. March 2004. "The Economic Benefits of Wilderness: Focus on Property Value and Enhancement." *Science and Policy Brief*, No. 2. Retrieved from: http://wilderness.org/sites/default/files/The-Economic-Benefits-of-Wilderness-With-a-Focus-on-Land-Value-Enhancement_low-res.pdf

Wilson, R. 2008, April 13. "Tellico Dam Still Generating Debate." *Knoxville News Sentinel*. Retrieved from: http://www.knoxnews.com/news/2008/apr/13/tellico-dam-still-generating-debate/

Wood, Harold. 2013. "Quotations from John Muir." Sierra Club. Retrieved from: http://www.sierraclub.org/john_muir_exhibit/writings/favorite_quotations.aspx

Wright, Justin P. and Clive G. Jones, Alexander G. Flecker. 2002. "An ecosystem engineer, the beaver, increases species richness at the landscape scale." *Oecologia* 132 (1): 96–101. Retrieved from: http://www.caryinstitute.org/sites/default/files/public/reprints/Wright_et_al_2002_An_ecosystem_Oecologia_132_96–101.pdf

Yonk, Ryan M., Brian C. Steed, and Randy T Simmons. 2011. Public Lands and the Political, Economic, and Social Structure of Rural Western Counties. Center for Public Lands and Rural Economics. Retrieved from: http://www.usu.edu/ipe/wp-content/uploads/ipePublications/Public-Lands-and-the-Political-Economic-and-Social-Structure-of-Rural-Western-Counties.pdf

Yonk, Ryan M., Randy T Simmons, and Brian C. Steed. 2013. *Green vs. Green: The Political, Legal, and Administrative Pitfalls Facing Green Energy Production*. New York: Routledge.

Zimmerman, Corinne and Kim Cuddington. October 2007. "Ambiguous, Circular and Polysemous: Students' Definitions of the 'Balance of Nature' Metaphor." *Public Understanding of Science*. 16: 393–406.

Zybach, Robert. 2012. "Oregon's 2012 Wildfires: Predictable and Preventable," *Oregon Fish & Wildlife Journal*, 34 (4): 7–17. Retrieved from: http://www.nwmapsco.com/ZybachB/Articles/Oregon_Wildfires_2012/Zybach_2012c.pdf

Index

Note: Page numbers followed by an *"n"* indicate an endnote with additional information about that topic.

A

aboriginal buffer zones, 12–13
absurd results doctrine, 62–63
Ackerman, Bruce, 56
Adler, Jonathan, 28, 70, 103
African Serengeti, 13
Afton, Wyoming, reinstallation of hydroelectric plant, 88
Agriculture Department, 113–14
agriculture industry and Chesapeake Bay, 113–14, 114–15
air pollution
    overview, 71
    from coal burning, 51–52, 56
    federal legislation, 53–54
    history of, 51–52
    introducing cleaner fuels, 53
    lethal levels in populated areas, 51
    measurement of, 52, 55
    and motor vehicle emissions standards, 53, 55, 56, 57
    photochemical smog, 53
    See also *Massachusetts v. Environmental Protection Agency*; entries beginning with "Clean Air Act"
Air Pollution Control Act (1955), 53–54
Air Quality Act (1967), 54
air quality laws, 29, 53–54, 55, 56. See also entries beginning with "Clean Air Act"
Alexander, George, 47
Allin, Craig, 160, 173
alternatives
    analysis under NEPA, 86–87, 97
    avoidance and minimization instead of mitigation, 117

    Corps of Engineers extremes, 116–17
    incompatibility of CWA and NEPA, 116–17
    most environmentally friendly choice, 116
Amazon River Basin, 14
American chestnut tree, 15–16
American Forest Resource Council, 79
American Recovery and Reinvestment Act (2009), 99, 179
America's Red Rock Wilderness Act (2009), 161–64
Anders, William, 43
Anne Arundel County, Maryland, and Chesapeake Bay TMDL, 114
Apollo 8 space mission Earthrise photograph, 42–43
Areas of Critical Environmental Concern, 207–8
Argonaut, 182
Arizona, red squirrels in, 131–32
arsenic level standards, 120–21
Asian chestnut blight, 15–16
Aspinall, Wayne, 160, 203
Association for the Preservation of the Little Tennessee River, 136
Atlanta, Georgia, pollution of Chattahoochee River, 107
automobiles and air pollution, 53, 55
avoidance and minimization instead of mitigation, 117
"Aw, Wilderness!" (Stroll), 163

B

Babbitt, Bruce, 138
Babbitt, Secretary of Interior v. Sweet Home Chapter of Communities for a Greater Oregon, 138–39, 145
Badè, William F., 165
Baker, Howard, 137–38
balance of nature
    overview, 7–8, 217n2(Ch. 2)

271

and Muir, 155
and species, 129–32
and sustained yield, 211–13
See also Endangered Species Act
balance of nature ideology
overview, 2–3, 5–6, 50
and CWA, 108–9
and ecosystems, 15–16
and ESA, 139–40
and NEPA, 44, 77–78
and "old-growth" forests, 16–19, 220n55
and political ecology, 8–9
and preservation of nature, 19–20
untouched wilderness myth, 13–14
and Wilderness Act, 157–60, 172
Balance of Nature: Ecology's Enduring Myth (Kricher), 8–9
Barnes, A. James, 70
barred owl vs. northern spotted owl, 140–42, 233n53
Bartlett, Randy, 114
Bay Journal (newspaper), 114
Beachfront Management Act (South Carolina), 145
Bear, Dinah, 92
beavers, 129–30
Behan, Dick, 32–33
biocentric ideologies, 46–47
biodiversity, 132. *See also* ecosystems; Endangered Species Act
Bishop, Rob, 76, 190–91
bison populations, 12
Black Elk Wilderness, South Dakota, 171–72
Black Hills National Forest, South Dakota, 170–72
blackout of August, 2003, 175
BLM. *See* U.S. Bureau of Land Management
Bobzien, Craig, 170–71
Bolick, Clint, 166–67
Boone and Crockett Club, 39
Botkin, Daniel, 7, 211–12
Bowen, Ronald, 114
Brademeyer, Brian, 171
buffer zones, 12–13
bureaucracies
overview, 204–5
agency-specific interpretations and collaborations, 96–97, 119–20
and CAA, 55–56
control loss process, 30–31
election-related problems, 31–32
Enlightenment model, 65
insufficient resources and additional time requirement, 92–94

interests of bureaucrats, 21–22
knowledge- and power-related problems, 64, 65–67
and NEPA projects, 88–89
pet projects disguised as mitigation, 87–89
and politics, 22–23
reciprocity, 80, 81, 91
and risks associated with innovation, 33–35
streamlining processes with correct incentives, 99–100
Wilderness Act and competition between agencies, 158–60
See also specific agencies and organizations
Bureau of Land Management. *See* U.S. Bureau of Land Management
burning as forest management, 16–19, 220n55
Bush, George W., 179

C
CAA. *See* Clean Air Act of 1970
Calico Mountains wild horse Complex, Nevada, 213
California
air pollution in Los Angeles, 53
landfill in southern California, 73
Muir Woods and Miwok Indians, 155
O'Shaughnessy Dam, Tuolumne River, 40–42
San Francisco, 41–42
Union Oil Company oil spill in Santa Barbara Channel, 25, 103–4
Canada lynx, 88, 130
Canyonlands National Park, Utah, 209–10
Carmichael family, Texas, 109–10
Carson, Rachel, 8, 25–26
Carter, Dick, 189
categorical exclusions under NEPA, 75, 76
cemeteries in Utah, 205–6
Center for Biological Diversity, 23–24, 89
CEQ (Council on Environmental Quality), 75, 76, 80
Chaffin, Gerald, 199, 206–7
Chase, Alston, 34–35
Chattahoochee River, Atlanta's pollution of, 107
Chesapeake Bay TMDL, 113–15
Chevron Defense, 62, 63
Chevron U.S.A., Inc. v. National Resources Defense Council, Inc., 62–63, 138
Chicken, Alaska, EPA enforcement of CWA in, 124
Christensen, Norman, 3
Chu, Steven, 180, 183
Clark, William, 9
Clawson, Marion, 208

Clean Air Act (1963), 54, 55
Clean Air Act of 1970 (CAA), 51–71
  overview, 71
  authors' suggestions for marginal changes, 194–95
  bureaucratic approach, 55–56
  and coal industry, 69–70
  decline in pollutants prior to, 29
  effect of *Massachusetts v. EPA*, 63, 64
  on emissions regulations, 63
  and EPA regulation of greenhouse gas emissions, 58
  new source performance standards, 55
  nonattainment areas, 55, 71
  passage of, 44–45, 55–56
  and PSD, 56
  and rent-seeking behaviors, 68–70
  Supreme Court interpretation of, 60–61
Clean Air Act of 1970 Amendments (1977), 56–57, 70
Clean Air Act of 1970 Amendments (1990), 57–58
Clean Coal Dirty Air (Ackerman and Hassler), 56
Clean Water Act of 1972 (CWA), 103–26
  overview, 125–26
  alternative incompatibility with NEPA, 116–17
  amendments to, 107–8
  arbitrary decisions by unqualified staff, 117–18
  and arsenic level standards, 120–21
  authors' suggestions for marginal changes, 192
  and balance of nature, 108–9
  and Chesapeake Bay TMDL, 113–15
  common law replacement, 109–11
  and Cuyahoga River fire, 25, 28–29, 103, 121
  enforcement of, 122–24, 125
  and funding availability, 125
  impetus for, 103–4
  LEDPA focus on avoidance and minimization instead of mitigation, 117
  maximizing mitigation, 119–20
  mitigation requests, 108
  point-source and nonpoint-source water pollutants, 105–6, 111–12
  race to the bottom created by, 124
  redundancy with NEPA, 115–16
  regulatory nightmares, 115–20
  remediation of impaired water, 107, 112
  restoration project impacts, 118–19
  and reversal of Supreme Court decision in *Illinois v. Milwaukee*, 111
  states' responsibilities, 106–7
  taxpayer costs, 115–16
  and TMDL, 111–15

Clean Water Act Amendments (1973), 45
Clean Water Act Amendments (1977), 107
Clean Water Act Amendments (1987), 107–8
Clements, Frederic Edward, 7
Cleveland, Ohio, Cuyahoga River fire, 25, 28–29, 103, 121
climate change
  EPA decision on whether greenhouse gas emissions contribute to, 57–59, 60–61
  and ESA, 139, 149, 232n40, 232n44
  global warming, 59
  as human imprint in remote places, 19
  and mountain pine beetle, 170
coal burning and air pollution, 51–52, 56
coal industry and CAA, 69–70, 71
Codevilla, Angelo, 64, 65–66
Cole, David N., 173
common law replacement for CWA, 109–11
Common School Fund, Oregon, 127–28
Congressional oversight with control loss, 29–31
Congressional Research Service on TMDL system, 112
"Conservation Economics" (Leopold), 197–98
conservationists
  and national parks, 39–40
  and trade in endangered species, 133
  utilitarian vision vs. preservationism, 40–42
  wealth and prominence in early societies, 39–40
  See also specific conservationists
conservation rental contracts, 193
constitutional takings, 144–46, 234n71
control loss process, 29–31
Convention on International Trade in Endangered Species of Wild Fauna and Flora (1973), 133
Corbett, Tom, 71
Corps of Engineers. *See* U.S. Army Corps of Engineers
cottonwoods and elk management, 34
Coulter, Michael, 96
Council on Environmental Quality (CEQ), 75, 76, 80
Cronon, William
  on balance of nature ideology, 7
  on concept of wilderness, 157
  on conservationists, 39
  on Tuolumne River dam at Hetch Hetchy, 41
  on wilderness, 37
  on Zahniser, 156
crown fires, 18–19, 167–68, 220n55
Cuyahoga River fire, Cleveland, Ohio, 25, 28–29, 103, 121
CWA. *See* Clean Water Act of 1972

CYA (cover your ass) mindset and litigation prevention, 94–96

D
Dahl, Robert A., 5
Darling, Jay, 42
data, 30–31, 33, 83
Davis, Peter, 110
DDT (dichlorodiphenyltrichloroethane), 25–27, 103, 150
dead zones in Chesapeake Bay, 113
deep ecology, 46–47
degradation clause of FLPMA, 204, 214
delay tactics. *See* environmental political entrepreneurs and NEPA
democracy in action, 22
designs and designers, 187–89
detailed statement under NEPA, 75
Dicks, Norm, 133–34
Dobyns, Henry, 11
DOE. *See* U.S. Department of Energy
Dolan v. City of Tigard (Oregon), 146
Donora, Pennsylvania, 51
DOT (U.S. Department of Transportation), 62
Draft Chesapeake Bay TMDL, 114
driveway repair project for inholding property owner, 92–93, 97–98
Durbin, Dick, 162

E
Earth Day, 43
Earth First!, 47–48
Earthrise photograph, 42–43
Ecodefense (Foreman), 47
ecological balance, 8
ecologists, 1–2, 7, 8
ecology. *See* balance of nature ideology; political ecology
economic benefits of wilderness, 165
ecosystems
　overview, 5–6, 15, 152
　African Serengeti, 13
　and environmental impact predictions, 77
　myths about, 15–16
　outbreak populations, 12
　and plant diseases brought by Europeans, 15–16
　recognition of humans as components of, 157
　wilderness management for preservation of, 167–69, 173
Ehrlich, Paul, 27–28
EISs. *See* environmental impact statements
elk, numbers of, 10, 14, 31, 34

Elliot State Forest and Common School Fund, 127–28
Elliott, Francis, 127–28
El Paso Corporation, 89
Emerson, Ralph Waldo, 165
emissions, greenhouse gas, 57–63, 69. *See also* motor vehicle emissions
endangered species, 15, 128, 134–35, 147
Endangered Species Act of 1973 (ESA), 127–52
　overview, 129, 133–36, 152–53
　authors' suggestions for marginal changes, 193–94
　Babbitt v. Sweet Home, 138–39, 145
　and climate change, 139, 232n40, 232n44
　and Elliott State Forest, Oregon, 127–28
　history of, 132–33
　incentive structure, 23
　June sucker fish example, 83–85
　lawsuits related to, 23–24
　passage of, 45
　property owners' responses, 128–29
　species and balance of nature, 129–32
　taxpayer costs, 149
　TVA v. Hill, 136–39, 232n34
Endangered Species Act's failure and unanticipated consequences, 139–52
　balance of nature ideology, 139–40
　ineffectiveness, 150–52
　killing barred owl in favor of northern spotted owl, 140–42, 233n53
　political entrepreneur opportunities, 147–49
　preemptive habitat destruction, 128–29, 142–44, 146
　shoot, shovel, and shut up method for removing endangered species, 128, 147
　uncompensated loss of use vs. constitutional takings, 144–46, 234n71
Endangered Species Act Amendments (1982), 135
Endangered Species Committee "God Squad," 137–38
Endangered Species Conservation Act (1969), 133
Endangered Species Preservation Act (1966), 133
endemic species, 130
Energy Department. *See* U.S. Department of Energy
energy industry and NEPA, 77
Energy Policy Act (2005)
　overview, 175–76
　authors' suggestions for marginal changes, 194
　Loan Guarantee Program, 178–79, 180
　See also Solyndra
Engelskirger, Carson, 170
entrepreneur, defining, 4
Environ Defenders, 78

Environmental Defense Fund, 144
environmental disasters
	Cuyahoga River fire, 25, 28–29, 103, 121
	DDT, 25–26
	and fight against industrial activity, 24–28
	as reason/excuse for federal regulations, 28–29
environmental impact statements (EISs)
	overview, 43–44, 76
	bogus mitigation recommendations, 89
	controversy as trigger for, 91–92
	litigation-proofing, 95
	multi-year study on impacts to bird species, 81
	and NEPA, 75
environmentalists, 24–25, 27–28, 82
environmental movement
	overview, 46
	and Apollo 8 space mission Earthrise photograph, 42–43
	biodiversity and ecological integrity focus, 46–50
	and disasters related to industrial activity, 24–28
	and federal legislation, 45–46
	and public meetings, 21
	Sagebrush Rebellion, 48, 207
	and utilitarian vision vs. preservationism, 41
	wise-use-related, 49–50
	See also specific organizations
environmental myths, 14–19
	ecosystem myths, 15–16
	old-growth forest myths, 16–19, 220*n*55
	untouched wilderness, 13–14
	See also balance of nature
environmental policies
	overview, 1, 44–45
	and balance of nature belief, 8–9
	ecologists' failure to make recommendations, 1–2
	and ecology of politics, 68–70
	foundations of, 42
	political entrepreneurs' use of, 4
environmental political entrepreneurs, 21–35
	overview, 35
	Bishop as, 190–91
	and Chesapeake Bay TMDL, 115
	and Congressional oversight with control loss, 29–31
	and costs of risky choices, 33–35
	CWA-related issues, 106–7, 108, 122–24, 126
	and ESA, 147–49
	and fight against industrial activity, 24–28
	and Keystone XL Pipeline, 196–97
	manipulating laws, 196

and mountain pine beetle, 170–72
Muir as, 155–56
and planning horizons, 32–33
and politics, 21–22, 31–32
solar energy zones on public lands, 3–4
state vs. national level for political battles, 28–29, 122
strategies, 4, 22–23
and Wilderness Act, 158–60
See also Solyndra
environmental political entrepreneurs and NEPA
	overview, 74, 78–79
	agency personnel bias-based obstruction, 98
	agency-specific NEPA interpretations, 96–97
	controversy as trigger for EIS instead less expensive reports, 91–92
	costs borne by taxpayers, 76–77, 78–79, 87, 91, 92, 95, 96
	cover your ass mindset, 94–96
	detailed analysis requests, 76, 80–82
	insufficient alternatives analysis, 86–87
	insufficient resources and additional time requirement, 92–94
	June sucker fish example, 83–85
	mitigation through fulfillment of pet projects, 87–89
	obscure laws and regulations roadblocks, 97–98
	overstating likely impacts, 89–91
	renewable energy projects as positive incentive for fast reviews, 99–100
	scoping issue requests late in the process, 83–85
	tactic overview, 79–80
environmental protection
	common law roots, 106, 109
	by federal government, 44–46
	FLPMA focus on, 207–8
	at state level, 29
	See also Clean Air Act; Clean Water Act; Endangered Species Act; Wilderness Act
Environmental Protection Agency (EPA)
	bureaucratic biases and WYSIATI thinking, 65–67
	creation of, 44
	and CWA enforcement, 122–24
	DDT ban, 26
	and greenhouse gas emissions, 57–63, 69
	insufficient resources/delay tactics, 93–94
	and NAAQS, 55
	National Evaluation of the CWA, 112
	and NPDES, 104–5, 106, 107
	Office of General Counsel and EIS length, 82

secret science for justifying CAA rule changes, 194–95
states' annual report on water quality, 106
Tailoring Rule, 62–63, 67
and TMDLs, 112
See also Massachusetts v. Environmental Protection Agency; motor vehicle emissions
Environmental Protection Agency, Utility Air Regulatory Group v., 63
EPA. *See* Environmental Protection Agency
equilibrium ecology, 2–3. *See also* balance of nature
ESA. *See* Endangered Species Act of 1973
ethanol Renewable Fuel Standard, 195

F

Fairfax County, Virginia, and Chesapeake Bay TMDL, 114
federal government
    assessment and control of potential environmental impacts, 43–44, 73
    shift from politics to government programs, 31–32
    state vs. national level for political battles, 28–29
    wise-use vs. regulations of, 49–50
    See also bureaucracies; *specific agencies and legislation*
Federal Housing Authority, 95
federal land policy, 199–214
    and Chaffin home on federal land, 199, 206–7
    conservation push, 201–3
    history of public land management, 200–201
    See also Federal Land Policy and Management Act of 1976
Federal Land Policy and Management Act of 1976 (FLPMA)
    overview, 204
    and Chaffin home on federal land, 199, 206–7
    federal ownership of public land mandate, 205
    Fruit Heights cemetery, Utah, 205–6
    multiple use mandate, 207–10
    passage of, 203
    politics of nature, 45
    sustained-yield mandate, 211–14
federal recovery plan under ESA, 151
Federal Water Pollution Control Act (1972), 104. *See also* Clean Water Act of 1972
feedback loop and Clean Air Act, 68–70
fires as forest management, 16–19, 220n55
Fish and Wildlife. *See* U.S. Fish and Wildlife Service
fishing, open access, and decline in fish populations, 211–12
Flatt, Victor B., 105
Florida, fish kill from food-processing plants' discharge, 104
FLPMA. *See* Federal Land Policy and Management Act of 1976
forbs, 18
Foreman, Dave, 47–48
forest management, burning as, 16–19, 220n55
Forest Service. *See* U.S. Forest Service
1491: New Revelations of the Americas Before Columbus (Mann), 11–12
Frederick, Maryland, and Chesapeake Bay TMDL, 114
Fristik, Richard, 88–89
Frome, Michael, 165
Fruit Heights cemetery, Utah, 205–6
Fruit Heights Lands Conveyance Act, 206
fuel economy standards, 62

G

Gamble, Rachel B., 164
GAO (U.S. Government Accountability Office), 121–22, 123
Garfield, James R., 41–42
Gates, Paul Wallace, 202
Geist, Valerius, 12
General Land Office, 200–201
General Land Reform Act (1891), 202
George Kaiser Family Foundation, 182
Georgia, Chattahoochee River pollution, 107
Gerard, David, 159
global warming. *See* climate change
Goklany, Indur, 52
Graham, Michael, 211
gray wolves, 130–31
grazing rights and site-specific information, 33
Greater Canyonlands region, Utah, 210
"Great Smog" in London, 51
Greenberg, Ken, 107
greenhouse gas emissions, 57–63, 69. *See also* motor vehicle emissions
Greenstone, Michael, 71
grizzly bears, 130–31
Gronet, Christian "Chris," 177, 183
ground fires, 17, 18
Guthrie, Woody, 202–3

H

habitat assessment, 193
habitat destruction method for preventing endangered species, 128–29, 142–44, 146

habitat modification as harm to a species under ESA, 138–39
Hassler, William T., 56
Hastings, Doc, 78–79, 149, 150
Hawkins, George, 114
Haws, J. Matthew, 86, 95, 96
Hayek, Friedrich, 64, 67
Hayes, Dennis, 43
Herbert, Gary, 169
HEW (U.S. Department of Health, Education, and Welfare), 54
Hill et al., Tennessee Valley Authority v., 136–39, 232n34
Hinchey, Maurice, 162
Hitchcock, Ethan A., 41
Hoffinger, Fran, 96
Hoffman, Jeff, 27
Homestead Act (1862), 201, 205
Hornaday, William Temple, 40
HUD (U.S. Department of Housing and Urban Development), 95
hunting for sustenance, 40
Hutcheson Memorial Forest, New Jersey, 17

I
Illinois v. Milwaukee, 110–11
India, 27
individuals, 7, 21, 22
industrialization and air pollution, 52
Industrial Revolution, 52, 201–2
information-gathering challenges, 30–31, 33, 83
Innes, Robert, 142
innovation, risks related to, 33–35
interest groups
 and Clean Air Act Amendments, 56–57
 and Clean Air Act–related rent seeking, 68–70
 competition between, 32
 and multiple use land management, 208
 and NEPA litigation, 95–96
 See also environmental movement; *specific groups*
Interior Department, 32, 138–39, 159, 201
International Mountain Bicycling Association, 163
interspecies conflict and U.S. Fish and Wildlife Service, 140–42
Isakowitz, Steve, 180

J
Jackson, Lisa, 123–24
Johnson, David B., 30–31
Johnson, Lyndon B., 156
June sucker fish, 83–85

K
Kabat, Geoffrey, 66
Kahneman, Daniel, 65–66, 67
Kaiser Ventures, Inc., 73
Kareiva, Peter, 5
Kay, Charles, 12
Keigley, Richard B., 34–35, 222n47
Kennedy, John F., 203
keystone species, 129–30
Keystone XL Pipeline, 196
Kingdon, John, 4
Kitt Peak, Arizona, 131
Klingle, Matthew, 40
knowledge economy, 64–67
Kricher, John, 8–9

L
Lacey Act (1900), 132
Laitos, Jan G., 164
Land and Water Conservation Act (1964), 132
Land and Water Conservation Fund, 28, 132–33
landfill in southern California, 73
Landres, Peter, 157, 167, 173
LaVera, Damon, 183
lawsuits
 overview, 45–46
 absurd results doctrine, 62–63
 avoidance/prevention efforts, 82, 89, 94–96, 169
 and ESA, 153
 establishing standing for bringing suit, 58–60
 Lucas v. South Carolina Coastal Council, 145
 as obstructionist tactic, 78, 79
 politics of nature, 45–46
 Sierra Club vs. EPA on air quality enforcement, 56
 Sierra Club vs. U.S. Forest Service, 48
 SUWA's WSA protection tactics, 163–64, 209
 Texas family's injunction against Arkansas for sewage, 109–10
 and WildEarth Guardians mission, 165–66
 See also U.S. Supreme Court
lead pollution, 51, 55
Lean, Geoffrey, 51
least environmentally damaging practicable alternative (LEDPA), 116, 117
Leavitt, Michael O., 189
LEDPA (least environmentally damaging practicable alternative), 116, 117
"Legal Handbook for Environmental Activists" (Environ Defenders), 78
legislative oversight with control loss, 29–31
Leopold, Aldo, 42, 197–98

Lewis, Meriwether, 9
litigation. *See* lawsuits
Living Wilderness magazine (Wilderness Society), 164–65
Loan Guarantee Program of Energy Policy Act, 178–79, 180
lobbying costs, 28–29
logistic population, 211–12
London, England, 51–52
Los Angeles, California, air pollution, 53
Love Canal, New York, industrial waste buried in, 25
Lucas v. South Carolina Coastal Council, 145
Lueck, Dean, 143
Lujan v. National Wildlife Federation, 59–60
Luther, Linda, 96–97

M
malarial control with DDT, 26–27
Mandelker, Daniel R., 77
Mann, Charles C., 11–12
Margro, Thomas, 82
Marshall, Bob, 160
Maryland and Chesapeake Bay TMDL, 114
Massachusetts, global warming effects in, 59
Massachusetts v. Environmental Protection Agency, 57–69
  overview, 57–58
  and Clean Air Act of 1970 interpretation, 63
  effect on EPA, 65–67
  establishing standing, 58–60
  events leading to Supreme Court Hearing of, 69
  and greenhouse gas emissions, 57–63, 69
  knowledge economy, 64–67
  Supreme Court's decision, 60–61
  as war on coal and jobs, 71
Maximum Sustainable Yield, 211–12
McIntosh, Heidi, 210
Michael, J. A., 143
Migratory Bird Conservation Act (1929), 132
Migratory Bird Treat Act (1918), 97–98, 132
Milwaukee, Illinois v., 110–11
Milwaukee, Wisconsin, polluting Lake Michigan, 110–11
minimization and avoidance instead of mitigation, 117
Mining and Minerals Policy Act (1970), 208
mitigation
  overview, 88–89, 118–19
  and CWA projects, 108
  and environmental political entrepreneurs, 87–89, 119–20
  and funding availability, 125

LEDPA focus on avoidance and minimization instead of mitigation, 117
  maximizing under CWA, 119–20
  and NEPA projects, 87–89
  requiring more from wealthy petitioners, 98
Miwok Indians, California, 155
moose, numbers of, 10
moral obligation to maintain wilderness, 38–39, 164–66
Morriss, Andrew P., 56–57, 68–69
motor vehicle emissions
  EPA responsibility for regulation of, 59, 61–62
  standards, 53, 55, 56, 57
  Tailoring Rule, 62–63, 67
mountain pine beetle, 170–72
Mount Graham, Arizona, 131–32
Mueller, John E., 22
Muir, John, 38, 39, 40–42, 155–56, 165
Muir Woods, California, 155
Müller, Paul, 26
Multiple Use and Sustainable Yield Act (1960), 207
multiple use mandate of FLPMA, 207–10
Muskie, Edmund, 54, 70
myths. *See* environmental myths

N
NAAQS (National Ambient Air Quality Standards), 55
Naess, Arne, 46
Nash, Roderick, 37
National Ambient Air Quality Standards (NAAQS), 55
National Environmental Policy Act of 1969 (NEPA), 73–101
  overview, 73–75, 100–101
  alternative incompatibility with CWA, 116–17
  authors' suggestions for marginal changes, 192–93
  balance of nature assumption, 44, 77–78
  cost of assessments and EISs, 76
  and Council on Environmental Quality, 75, 76, 80
  landfill in southern California example, 73
  passage of, 25, 43–44
  as procedural law, 75–76
  redundancy with CWA, 115–16
  and Tellico Dam, Tennessee, 136, 147–48
  See also environmental political entrepreneurs and NEPA
National Evaluation of the CWA (EPA), 112
national forests
  Black Hills National Forest, South Dakota, 170–72
  central planning for, 32–33

National Park Service takeover, 159
public enjoyment of, 202
roadless area reviews, 48
utilitarian vision, 40
See also U.S. Forest Service (USFS)
national monument status, 210
National Oceanic and Atmospheric Administration (NOAA), 134, 152, 153. *See also* Endangered Species Act
national parks, 39–40. *See also* U.S. National Park Service
National Pollution Discharge Elimination System (NPDES), 104–5, 106, 107
National Resources Defense Council, Inc., Chevron U.S.A., Inc. v., 62–63, 138
National Wilderness Preservation System, 156, 160–61, 190
National Wildlife Federation, Lujan v., 59–60
Native Americans
  aboriginal buffer zones, 12–13
  and agriculture, 11
  evictions from national parks, 155
  Muir's views on, 39
  and "old-growth" forests, 16–18
  and untouched wilderness myth, 13–14
  and white man's diseases, 11, 12
nature undisturbed. *See* balance of nature ideology
NEPA. *See* National Environmental Policy Act; National Environmental Policy Act of 1969
Neumann, Thomas W., 11–12
New Jersey, Rutgers University's Hutcheson Memorial Forest, 17
news media for delay tactics, 90
new source performance standards under CAA, 55
New York pulp mill contaminating a creek, 110
NIMBY (Not in My Backyard), 78. *See also* environmental political entrepreneurs
Nixon, Richard M., 44, 54–55, 74
NOAA (National Oceanic and Atmospheric Administration), 134, 152, 153. *See also* Endangered Species Act
no-action alternatives under NEPA, 86
nonattainment areas under CAA, 55, 71
nonpoint-source water pollutants, 105–6, 111–12
North American wolverine, 130
North Dakota and CWA enforcement, 124
northern spotted owl, 138–39, 140–42, 233n53
Not in My Backyard (NIMBY), 78. *See also* environmental political entrepreneurs
NPDES (National Pollution Discharge Elimination System), 104–5, 106, 107

O
Obama, Barack
  Chesapeake Bay restoration and protection order, 113
  inaugural celebration, 202–3
  and Solyndra, 176–77, 179–80, 181–82, 184–85
obscure laws and regulations roadblocks, 97–98
ocular reconnaissance method for determining range quality, 34
off-road vehicles (ORVs), 32, 163–64
Ogden, Peter Skene, 10
Ohio, Cuyahoga River fire, 25, 28–29, 103, 121
oil spills via railcars vs. pipelines, 197
old-growth forest myths, 16–19, 220n55
old-growth forests, 127–28
One Third of the Nation's Land (Public Land Law Review Commission), 203
Oregon
  Babbitt v. Sweet Home, 138–39, 145
  Dolan v. City of Tigard, 146
  Elliot State Forest and Common School Fund, 127–28
  Ruby Pipeline between Wyoming and, 89
ORVs (off-road vehicles), 32, 163–64
outbreak populations, 12

P
paralysis by analysis and NEPA, 80
Pederson, William F., 68
Pennsylvania, 51, 71, 114–15
permit requirements
  for greenhouse gases, 62–63
  political entrepreneurs' manipulation of, 119–20
  for pollutant discharge into national waterways, 104–5, 106, 107
  projects that affect water, 108
  Section 404 of CWA, 115–16
  See also environmental impact statements
pet projects and mitigation, 87–89
Phelan, James D., 41
photochemical smog, 53
Pielou, E.C., 16
pigeon populations, 11–12
Pinchot, Gifford, 40, 41
planning horizons, 32–33
Plater, Zygmunt, 136–37, 232n34
point-source water pollutants, 105–6, 111–12
political ecology, 7–20
  overview, 1–3, 5–6
  and balance of nature belief, 8–9
  and bureaucracies, 67
  and wildlife, 9–13
  See also balance of nature ideology

political entrepreneurship, 3, 4, 22–23, 190–91. *See also* environmental political entrepreneurs; National Environmental Policy Act
politics
　overview, 5, 21–22, 35
　allocating costs to one group and benefits to another, 166–67
　election-related problems, 31–32
　interests of politicians, 21, 22
　lobbying costs, 28–29
　misleading or slanted information, 30–31
　principles for redesigning environmentalism, 187–88
politics of nature, 37–50
　overview, 37, 50
　and Apollo 8 space mission, 42–43
　deep ecology, 46–48
　FLPMA, 45
　legislation and court battles, 45–46
　NEPA, 43–45
　RARE II, 48
　utilitarian vision vs. preservationism, 40–42
　wealth vs. poverty, 39–40
　wilderness cult/movement, 37–39
　wise-use movement, 40–42, 49–50
ponderosa pine forests, Arizona, 18
Population Bomb, The (Ehrlich), 27–28
population control, 27–28
poverty vs. wealth and conservation, 39–40
power outage of August, 2003, 175
preemptive habitat destruction, 128–29, 142–44, 146
"Preemptive Habitat Destruction Under the Endangered Species Act" (Lueck and Michael), 143
preservation of nature, 19–20
Prevention of Significant Deterioration (PSD) doctrine, 56
Profeta, Timothy, 158
property rights
　and common-law system, 109–11
　CWA erosion of, 106
　endangered species habitat destruction for retaining property rights, 128–29, 142–44, 146
　and ESA, 134, 135–36, 138–39
　wilderness designations vs., 166–67
　and wise-use movement, 40–42, 49–50
　See also takings
PSD (Prevention of Significant Deterioration) doctrine, 56
Public Land Law Review Commission, 203

Q
Quiet Crisis, The (Udall), 28

R
Raker Act (1913), 42
Rambunctious Garden (Marris), 19–20
Randall, Gretchen, 116
RARE II (roadless area reviews by U.S. Forest Service), 48
rare species, 130
Raynolds, F. W., 12–13
Reagan, Ronald, 48, 122, 207
"Rebuilding the Ark" (Environmental Defense Fund), 144
recovery plan under ESA, 151
red-cockaded woodpecker, 128–29, 138–39, 143
red squirrels, 131
religious view of nature
　author's dismissal of, 187, 188
　Muir, 38, 39, 41–42, 155, 165
　political gridlock resulting from, 164–66
　Roosevelt, Theodore, 38–39
　SUWA, 210
renewable energy programs, 175–86
　overview, 185–86
　as incentive for fast NEPA reviews, 99–100
　taxpayer costs, 176, 177, 184, 185–86
　See also Energy Policy Act; Solyndra
rent seeking, 68–70, 195, 196
resiliency of ecosystems, 15
restoration and balance of nature ideology, 139–40
restoration project impacts under CWA, 118–19
Riccardi, Nathan D., 62
Ridenour, David, 150
rights, 49. *See also* property rights
right to bear arms, 49
Ringelmann chart for measuring air pollution, 52
risky choices, costs of, 33–35
river restoration project challenges, 118–19
roadless area reviews, 48
Rockwell, Norman, 22
Roosevelt, Theodore, 38–39, 155–56, 202
Rotman, Michael, 29
Ruby Pipeline between Wyoming and Oregon, 89
Russell, E. S., 211
Rutgers University's Hutcheson Memorial Forest, New Jersey, 17

S
Sagebrush Rebellion, 48, 207
Salt Creek Canyon, Utah, 209–10
Salter, Trevor, 99
Sand County Almanac, A (Leopold), 42
San Francisco, California, 41–42
Santorum, Rick, 148
Schullery, Paul, 9
Schuyler, Arent, 25

Schwarzenegger, Arnold, 183
science and politics, 1–2
science of ecology, 1–2
scoping period for NEPA documents, 83
Serengeti, African, 13
Seward, Lachlan, 179
Sexton, Ken, 66, 67
Shell Oil Company, 95–96
shoot, shovel, and shut up method for removing endangered species, 128, 147
Silent Spring (Carson), 8, 25–26
Sim, Kenneth J. (author), 50, 92–94
Simmons, Randy T, 50
Simmons, Randy T (author), 32, 34, 50, 92–94
Simpson, Mike, 148
site-specific information and grazing rights, 33
slickspot peppergrass, 148
Small Tracts Act, 207
Smillie, Susan M., 82
Smith, Robert J., 142–43
Snail Darter and the Dam, The (Plater), 136–37
snail darter fish battle, Tennessee, 136–39, 232n34
solar energy zones on public lands, 3–4
Solyndra, 176–85
   overview, 176–78
   failure of, 184–85
   federal funding process, 178–84
   House of Representatives Energy and Commerce Committee report on, 181–82
   investment rating, 182
   loan approval goes through, 180–81
   "Solyndra Failure, The" (House of Representatives Energy and Commerce Committee), 181–82
Soto, Hernando de, 12
South Carolina Beachfront Management Act, 145
South Dakota, Black Hills National Forest and mountain pine beetle, 170–72
Southern Utah Wilderness Alliance (SUWA)
   and America's Red Rock Wilderness Act, 161–62
   lawsuits filed by, 163–64, 209–10
   and Leavitt wilderness proposal, 189
   and multiple use management practices, 208–9
   SUWA v. Norton, 163–64, 209
   and Utah's attempt to shift federal land to the state, 169
special interests. *See* interest groups
spiked tree and logger injuries, 47
spirituality. *See* religious view of nature
state environmental protections, 29
   federal concerns about "race to the bottom," 104, 122

progress prior to CWA, 104, 121–22
of waterways, 104, 106, 109–11
state forest, Oregon, 127–28
state implementation plans (SIP) under CAA, 55, 57
state vs. national level for political battles, 28–29
steady-state (balance of nature) belief and solutions, 2–3
stimulus package (2009), 99
Stroll, Ted, 163
success of ESA, defining, 150–51, 152
Suckling, Kieran, 24
Supreme Court. *See* U.S. Supreme Court
sustained-yield mandate of FLPMA, 211–14
SUWA. *See* Southern Utah Wilderness Alliance
SUWA v. Norton, 163–64, 209
Swartz, Lucinda Low, 82
Sweet Home Chapter of Communities for a Greater Oregon, Babbitt, Secretary of Interior v., 138–39, 145

T
Tabb, William Murray, 91
Tailoring Rule, 62–63, 67
takings
   ESA definition, 135, 138, 232n36
   uncompensated loss of use vs. constitutional takings, 144–46, 234n71
taxpayer costs
   overview, 45
   allocating costs to one group and benefits to another, 166–67
   America's Red Rock Wilderness Act, 162
   CWA issues, 115–16
   ESA issues, 149
   of litigation process, 96
   NEPA issues, 76–77, 78–79, 87, 91, 92, 95, 96
   renewable energy programs, 176
   for restoration and protection of wilderness, 166
   Solyndra failure, 177, 184, 185–86
   Wilderness Act issues, 165, 167
   See also takings
Taylor Grazing Act (1934), 202
Taylor, Joseph E., III, 40
Tellico Dam, Little Tennessee River, Tennessee, 136–39, 147–48, 232n34
Tennessee Valley Authority v. Hill et al., 136–39, 232n34
Texas, 109–10, 143–44
"Thinking, Fast and Slow" (Kahneman), 65–66
"This Land Is Your Land" (Guthrie), 202–3
Thompson, W. F., 211
Thousand-Mile Walk to the Gulf, A (Muir), 165

282 | Index

Tigard, Oregon, 146
timber industry
    Elliot State Forest and Common School Fund, 127–28
    and ESA, 138–39, 143, 148
    and mountain pine beetle, 170–71
    and multiple use policy under FLPMA, 208
    spiked tree and logger injuries, 47
    and Wilderness Act, 157
Time Magazine, 103
TMDL (total maximum daily load), 111–15
Transfer of Public Lands Act (Utah), 169
Transportation Department, 62
trophic pyramid, 130
Tuolumne River dam, California, 40–42
Turner, Frederick Jackson, 38
Turner, James, 43, 47, 49

U
Udall, Stewart L., 28
uncompensated loss of use vs. constitutional takings, 144–46, 234n71
Union Oil Company oil spill in Santa Barbara Channel, 25, 103–4
unnecessary or undue degradation clause of FLPMA, 204, 214
U.S. Army Corps of Engineers
    alternative analysis extremes, 87, 116–17
    arbitrary decisions by unqualified staff, 117–18
    and mitigation under CWA, 119–20
    and NEPA documents, 192–93
    paternalistic protection of resources, 89–90
U.S. Bureau of Land Management (BLM)
    and Chaffin home on federal land, 199, 206–7
    driveway repair project for inholding property owner, 92–93, 97–98
    and FLPMA, 45, 204, 209, 214
    grazing district management, 202
    Instruction Memorandum prioritizing wind and solar energy projects, 100
    as mediator of multiple use, 207–10
    and off-road vehicle use in WSA, 163–64
    and range conditions information, 33
    and solar energy zones on public lands, 3–4
    wild horse management, 213–14
    See also Federal Land Policy and Management Act of 1976
U.S. Department of Agriculture (USDA), 113–14
U.S. Department of Energy (DOE)
    and Energy Policy Act's Loan Guarantee Program, 176, 178, 179, 180–84
    on time required for NEPA assessments and EISs, 76
U.S. Department of Health, Education, and Welfare (HEW), 54
U.S. Department of Housing and Urban Development (HUD), 95
U.S. Department of the Interior (DOI), 32, 138–39, 159, 201
U.S. Department of Transportation (DOT), 62
"Use of Knowledge in Society, The" (Hayek), 64
U.S. Fish and Wildlife Service
    overview, 152
    and EISs, 81, 84–85
    mitigation demands under CWA Section 404, 120
    northern spotted owl vs. barred owl, 140–42, 233n53
    quest for a balanced natural state, 140
    See also Endangered Species Act
U.S. Forest Service (USFS)
    competition with National Park Service, 158–60
    and FLPMA multiple use mandates, 204
    and Fruit Heights cemetery, 206
    lawsuits and administrative appeals, 48
    litigation effects, 96
    management plans, 31, 32–33
    mountain pine beetle control, 171
    NEPA projects and mitigation, 88
    ranchers' compensation for wolf predation, 147
    and roadless area reviews, 48
    and Small Tracts Act, 207
    utilitarian vision vs. preservationism, 40–42
    Wilderness Act competition with National Park Service, 158–60
    See also national forests
U.S. Geological Survey, 162
U.S. Government Accountability Office (GAO), 121–22, 123
U.S. National Park Service
    competition with Forest Service, 158–60
    and FLPMA multiple use mandates, 204
    and Keigley data and testimony before the U.S. House, 34–35, 222n47
    SUWA lawsuit against, 209–10
    and wilderness management, 167–69
    and Yellowstone National Park, 14, 31
U.S. Supreme Court
    Babbitt v. Sweet Home, 138–39, 145
    Chevron U.S.A., Inc. v. National Resources Defense Council, Inc., 62–63, 138
    Dolan v. City of Tigard (Oregon), 146
    Illinois v. Milwaukee decision and reversal, 110–11
    Lujan v. National Wildlife Federation, 59–60

SUWA v. Norton, 163–64, 209
TVA v. Hill, 136–39
Utility Air Regulatory Group v. EPA, 63
See also *Massachusetts v. Environmental Protection Agency*
Utah
　Fruit Heights cemetery, 205–6
　plan to shift federally-controlled lands to state control, 169
　proposals to classify wilderness areas, 189–91
　SUWA, 161–64
　Utah Lake and Provo River, Utah, 83–85
　Washington County Growth and Conservation Act of 2008, 189–90
Utah Wilderness Association, 189
Utility Air Regulatory Group v. Environmental Protection Agency, 63

V

Vines, Jim, 78, 95, 101
Virginia and Chesapeake Bay TMDL, 114
"Virginia Deer and Intertribal Buffer Zones . . ." (Hickerson), 12

W

Washington County Growth and Conservation Act of 2008 (Utah), 189–90
water quality standards, 105, 107, 112. *See also* Clean Water Act of 1972
waterway protections under CWA, 115–20
wealth vs. poverty and conservation, 39–40
Western Lands Project, 3–4
Western Watersheds Project, 89
West, Oswald, 127–28
wetland protections under CWA, 115–20, 118–19
whole economy modeling, 194
"Why the Clean Air Act Works Badly" (Pederson), 68
Wild and Scenic Rivers Act, 28, 190
WildEarth Guardians, 165–66
wilderness
　overview, 37
　proposals for management, 162–63, 188–91, 208–9
　support for, 164–66
　untouched wilderness myth, 13–14
　wilderness cult/movement, 37–39
　See also politics of nature
Wilderness Act (1964), 155–74
　overview, 172–74
　authors' suggestions for marginal changes, 194
　and balance of nature ideology, 157–60, 172
　management model, 158, 167–69, 171–72
　moral reasons vs. economic costs for wilderness designations, 164–67
　and mountain pine beetle, 170–72
　origin and intent, 155–57
　preservationist language, 163, 167, 169, 173
　SUWA case study, 160–64
　taxpayer costs, 165, 167
　See also Wilderness Study Areas
"Wilderness Management 101" (Forest Service), 162–63
Wilderness Society, 42, 156, 164–65
Wilderness Study Areas (WSAs)
　classification change to wilderness, 189, 190
　inappropriate use of, 164, 209
　regulations applied to, 163–64
　sunset provision suggestion, 194
　SUWA v. Norton challenge to management of, 163–64, 209
Wilderness Watch, 168–69
Wild Free-Roaming Horses and Burros Act, 213
wild horse management, 213–14
wildlife, 9–13. *See also* Endangered Species Act of 1973; *specific animals*
Wilson, Woodrow, 42
wise-use movement, 40–42, 49–50
World Wildlife Fund, 8
WSAs. *See* Wilderness Study Areas
Wyoming, Ruby Pipeline between Oregon and, 89
WYSIATI (what you see is all there is), 65, 66

Y

Yellowstone National Park, 10, 14, 31, 34, 202
Yonk, Ryan M. (author), 50, 92–94
Yosemite National Park, 40–42, 155–56

Z

Zahniser, Howard, 156, 173

Acknowledgments

THIS BOOK IS a collaboration among principal authors and the student research associates who work at Strata. The idea for this book originated in 2002 with David J. Theroux, President of the Independent Institute, when he, political scientist Randy Simmons, and wildlife ecologist Charles Kay began talking about the intersection of economics, politics, and ecology. A draft manuscript was written at the time but then because of unforeseen health and other problems sat on the shelf for ten years. In this final book, the core ideas are all here and embedded in all of the chapters. The most explicit use of that manuscript is in Chapter 1 where we explain and explore the ideas that David, Charles and Randy have called "political ecology." We are very grateful to David and his colleagues at each step of the book's development. In particular, we want to thank Independent's acquisitions director Roy M. Carlisle, publications director Gail Saari, and research director William F. Shughart II for their irreplaceable insights and assistance.

The students involved in this project were primarily economics and political science undergraduates who immersed themselves in the topics of political entrepreneurship and ecology. They wrote first drafts of many of the chapters and commented on each other's work and our own as we moved through the process of producing a final manuscript. The Strata student research associates who contributed greatly to our work and the topics they explored, researched, and wrote on are:

- Neal Mason wrote on the history of the environmental movement. His research is primarily incorporated in Chapter 2.

- Justine Larsen did a lot of the background research on the Federal Land Policy Management Act of 1976. Her research is incorporated in Chapter 2.
- Grant Patty researched and wrote the background documents on wilderness and the Wilderness Act of 1964 that we used to produce Chapter 8. He also contributed to Chapter 4, the chapter on the Clean Air Act.
- Lindsey McBride critiqued the penultimate draft of Chapter 8 and made significant contributions to the revisions that are incorporated into the final draft.
- Richard Criddle produced the first draft of what became Chapter 7, The Endangered Species Act. His analytical wit and rigor are evident in the final draft of that chapter.
- Josh DeFriez was the student researcher primarily in charge of gathering background information and writing an early draft of the chapter on the Clean Air Act of 1970 and its subsequent amendments. His work is incorporated into Chapter 4.
- Megan Hansen contributed to Chapter 4 (The Clean Air Act) and Chapter 9 (Renewable Energy Regulations).
- Nick Hilton and Megan Hansen produced the background materials we used to produce Chapter 9.
- Kayla Harris supervised and coordinated the research and writing done by the student research associates. She prodded, pulled, and directed them and edited early drafts of many of the chapters.

About the Authors

RANDY T SIMMONS is Senior Fellow at the Independent Institute, Professor of Economics and Director of the Institute of Political Economy at Utah State University's Jon M. Huntsman School of Business, and former Mayor of Providence, Utah.

Professor Simmons's books include the award-winning *Beyond Politics: The Roots of Government Failure*, *Aquanomics: Water Markets and the Environment* and *The Political Economy of Culture and Norms: Informal Solutions to the Commons Problem*. A contributing author to various volumes such as *Re-Thinking Green: Alternatives to Environmental Bureaucracy*, he is the author of scholarly articles that have appeared in numerous journals, and his popular articles have been published in newspapers and magazines across the United States.

He received his Ph.D. in political science from the University of Oregon, and he is a member of the Board of Directors of the Utah League of Cities and Towns and a member of the Utah Governor's Privatization Commission.

RYAN M. YONK is Research Fellow at the Independent Institute, Research Director for the Institute of Political Economy in the Department of Economics at Utah State University, and Research Assistant Professor of Economics at Utah State University. Professor Yonk is also the Executive Director of Strata Policy.

Professor Yonk's articles have appeared in such scholarly journals as *Public Choice*, *The Independent Review*, *Political Perspectives*, *Applied Research in Quality of Life*, and *Journal of Public and Municipal Finance*.

His books include *Direct Democracy in the United States: Petitioners as a Reflection of Society* (with Shauna Reilly), and *Green vs. Green: Conflicts in Environmental Priorities* (with Randy T Simmons and Brian Steed). He is also the author of policy reports for the U.S. Department of Energy, Center for Public Lands and Rural Economics at Utah State University, Utah Taxpayers Foundation, Utah Public Service Commission, the Mercatus Center, and Strata Policy.

He received his Ph.D. in political science from Georgia State University.

KENNETH J. SIM is Director of the Reliable Energy Education Network and a former Analyst with Strata Policy. He specializes in environmental policy and has particular experience working with the National Environmental Policy Act and Clean Water Act. He received both a Master's degree in geography and Bachelor's degree in environmental studies from Utah State University, and he is the author of policy reports for the Mercatus Center, Southern Utah University, and Strata Policy.

Independent Studies in Political Economy

THE ACADEMY IN CRISIS | *Ed. by John W. Sommer*
AGAINST LEVIATHAN | *Robert Higgs*
AMERICAN HEALTH CARE | *Ed. by Roger D. Feldman*
ANARCHY AND THE LAW | *Ed. by Edward P. Stringham*
ANTITRUST AND MONOPOLY | *D. T. Armentano*
AQUANOMICS | *Ed. by B. Delworth Gardner & Randy T. Simmons*
ARMS, POLITICS, AND THE ECONOMY | *Ed. by Robert Higgs*
A BETTER CHOICE | *John C. Goodman*
BEYOND POLITICS | *Randy T. Simmons*
BOOM AND BUST BANKING | *Ed. by David Beckworth*
CALIFORNIA DREAMING | *Lawrence J. McQuillan*
CAN TEACHERS OWN THEIR OWN SCHOOLS? | *Richard K. Vedder*
THE CHALLENGE OF LIBERTY | *Ed. by Robert Higgs & Carl P. Close*
THE CHE GUEVARA MYTH AND THE FUTURE OF LIBERTY | *Alvaro Vargas Llosa*
CHOICE | *Robert P. Murphy*
THE CIVILIAN AND THE MILITARY | *Arthur A. Ekirch, Jr.*
CRISIS AND LEVIATHAN, 25TH ANNIVERSARY EDITION | *Robert Higgs*
CUTTING GREEN TAPE | *Ed. by Richard L. Stroup & Roger E. Meiners*
THE DECLINE OF AMERICAN LIBERALISM | *Arthur A. Ekirch, Jr.*
DELUSIONS OF POWER | *Robert Higgs*
DEPRESSION, WAR, AND COLD WAR | *Robert Higgs*
THE DIVERSITY MYTH | *David O. Sacks & Peter A. Thiel*
DRUG WAR CRIMES | *Jeffrey A. Miron*
ELECTRIC CHOICES | *Ed. by Andrew N. Kleit*
THE EMPIRE HAS NO CLOTHES | *Ivan Eland*
THE ENTERPRISE OF LAW | *Bruce L. Benson*
ENTREPRENEURIAL ECONOMICS | *Ed. by Alexander Tabarrok*
FINANCING FAILURE | *Vern McKinley*
THE FOUNDERS' SECOND AMENDMENT | *Stephen P. Halbrook*
GLOBAL CROSSINGS | *Alvaro Vargas Llosa*
GOOD MONEY | *George Selgin*
GUN CONTROL IN THE THIRD REICH | *Stephen P. Halbrook*
HAZARDOUS TO OUR HEALTH? | *Ed. by Robert Higgs*
HOT TALK, COLD SCIENCE | *S. Fred Singer*
HOUSING AMERICA | *Ed. by Randall G. Holcombe & Benjamin Powell*
JUDGE AND JURY | *Eric Helland & Alexender Tabarrok*
LESSONS FROM THE POOR | *Ed. by Alvaro Vargas Llosa*
LIBERTY FOR LATIN AMERICA | *Alvaro Vargas Llosa*
LIBERTY FOR WOMEN | *Ed. by Wendy McElroy*
LIVING ECONOMICS | *Peter J. Boettke*

MAKING POOR NATIONS RICH | *Ed. by Benjamin Powell*
MARKET FAILURE OR SUCCESS | *Ed. by Tyler Cowen & Eric Crampton*
THE MIDAS PARADOX | *Scott Sumner*
MONEY AND THE NATION STATE | *Ed. by Kevin Dowd & Richard H. Timberlake, Jr.*
NATURE UNBOUND | *Randy T Simons, Ryan M. Yonk, and Kenneth J. Sim*
NEITHER LIBERTY NOR SAFETY | *Robert Higgs*
THE NEW HOLY WARS | *Robert H. Nelson*
NO WAR FOR OIL | *Ivan Eland*
OPPOSING THE CRUSADER STATE | *Ed. by Robert Higgs & Carl P. Close*
OUT OF WORK | *Richard K. Vedder & Lowell E. Gallaway*
PARTITIONING FOR PEACE | *Ivan Eland*
PATENT TROLLS | *William J. Watkins, Jr.*
PLOWSHARES AND PORK BARRELS | *E. C. Pasour, Jr. & Randal R. Rucker*
A POVERTY OF REASON | *Wilfred Beckerman*
THE POWER OF HABEAS CORPUS IN AMERICA | *Anthony Gregory*
PRICELESS | *John C. Goodman*
PROPERTY RIGHTS | *Ed. by Bruce L. Benson*
THE PURSUIT OF JUSTICE | *Ed. by Edward J. López*
RACE & LIBERTY IN AMERICA | *Ed. by Jonathan Bean*
RECARVING RUSHMORE | *Ivan Eland*
RECLAIMING THE AMERICAN REVOLUTION | *William J. Watkins, Jr.*
REGULATION AND THE REAGAN ERA | *Ed. by Roger E. Meiners & Bruce Yandle*
RESTORING FREE SPEECH AND LIBERTY ON CAMPUS | *Donald A. Downs*
RESURGENCE OF THE WARFARE STATE | *Robert Higgs*
RE-THINKING GREEN | *Ed. by Robert Higgs & Carl P. Close*
RISKY BUSINESS | *Ed. by Lawrence S. Powell*
SECURING CIVIL RIGHTS | *Stephen P. Halbrook*
STRANGE BREW | *Douglas Glen Whitman*
STREET SMART | *Ed. by Gabriel Roth*
TAKING A STAND | *Robert Higgs*
TAXING CHOICE | *Ed. by William F. Shughart, II*
THE TERRIBLE 10 | *Burton A. Abrams*
THAT EVERY MAN BE ARMED | *Stephen P. Halbrook*
TO SERVE AND PROTECT | *Bruce L. Benson*
VIETNAM RISING | *William Ratliff*
THE VOLUNTARY CITY | *Ed. by David T. Beito, Peter Gordon, & Alexander Tabarrok*
WINNERS, LOSERS & MICROSOFT | *Stan J. Liebowitz & Stephen E. Margolis*
WRITING OFF IDEAS | *Randall G. Holcombe*

For further information:
510-632-1366 • orders@independent.org • http://www.independent.org/publications/books/